T0140024

Studies in Big Data

Volume 42

Series editor

Janusz Kacprzyk, Polish Academy of Sciences, Warsaw, Poland
e-mail: kacprzyk@ibspan.waw.pl

The series "Studies in Big Data" (SBD) publishes new developments and advances in the various areas of Big Data- quickly and with a high quality. The intent is to cover the theory, research, development, and applications of Big Data, as embedded in the fields of engineering, computer science, physics, economics and life sciences. The books of the series refer to the analysis and understanding of large, complex, and/or distributed data sets generated from recent digital sources coming from sensors or other physical instruments as well as simulations, crowd sourcing, social networks or other internet transactions, such as emails or video click streams and others. The series contains monographs, lecture notes and edited volumes in Big Data spanning the areas of computational intelligence including neural networks, evolutionary computation, soft computing, fuzzy systems, as well as artificial intelligence, data mining, modern statistics and operations research, as well as self-organizing systems. Of particular value to both the contributors and the readership are the short publication timeframe and the world-wide distribution, which enable both wide and rapid dissemination of research output.

More information about this series at http://www.springer.com/series/11970

Ali Emrouznejad · Vincent Charles
Editors

Big Data for the Greater Good

Editors
Ali Emrouznejad
Operations and Information Management, Aston Business School
Aston University
Birmingham
UK

Vincent Charles
Buckingham Business School
University of Buckingham
Buckingham
UK

ISSN 2197-6503 ISSN 2197-6511 (electronic)
Studies in Big Data
ISBN 978-3-030-06576-8 ISBN 978-3-319-93061-9 (eBook)
https://doi.org/10.1007/978-3-319-93061-9

Printed on acid-free paper

This Springer imprint is published by the registered company Springer International Publishing AG part of Springer Nature
The registered company address is: Gewerbestrasse 11, 6330 Cham, Switzerland

Preface

Today, individuals and organizations are changing the world with Big Data. Data has become a new source of immense economic and social value. Advances in data mining and analytics and the massive increase in computing power and data storage capacity have expanded, by orders of magnitude, the scope of information available to businesses, government, and individuals. Hence, the explosive growth in data and Big Data Analytics has set the next frontier for innovation, competition, productivity, and well-being in almost every sector of our society, from industry to academia and the government. From ROI calculations, revenue impact analyses, and customer insights, to earthquake prediction and environmental scanning, suicide and crime prevention, fraud detection, better medical treatments, and poverty reduction—this is a reminder of all the good that can come from Big Data. But driving knowledge and value from today's mountains of data also brings policies related to ethics, privacy, security, intellectual property, and liability to the forefront, as the main concern on the agenda of policy makers. To help Big Data deliver its societal promises will require revised principles of monitoring and transparency, thus new types of expertise and institutions.

The present book titled 'Big Data for the Greater Good' brings together some of the fascinating uses, thought-provoking changes, and biggest challenges that Big Data can convey to our society. Along theory and applications, the book compiles the authors' experiences so that these may be aggregated for a better understanding. This book should be of interest to both researchers in the field of Big Data and practitioners from various fields who intend to apply Big Data technologies to improve their strategic and operational decision-making process. Finally, we hope that this book is an invitation to more intensive reflection on Big Data as a source for the common and greater good.

The book is well organized in nine chapters, contributed by authors from all around the globe: Austria, Brazil, Czech Republic, Denmark, France, Germany, Greece, Italy, The Netherlands, UK, and the USA.

Chapter 1 provides an introduction to Big Data, as well as the role it currently plays and could further play in achieving outcomes for the 'Greater Good'. This chapter discusses the main literature on Big Data for the Greater Good for interested readers coming from different disciplines. Chapter 2 looks at the means that can be used to extract value from Big Data, and to this end, it explores the intersection between Big Data Analytics and Ethnography. The authors advance that the two approaches to analysing data can complement each other to provide a better sense of the realities of the contexts researched. As such, Big Data Analytics and Ethnography together can assist in the creation of practical solutions that yield a greater societal value.

Chapter 3 complements the first two chapters and provides a global overview of what Big Data is, uncovers its origins, as well as discusses the various definitions of the same, along with the technologies, analysis techniques, issues, challenges, and trends related to Big Data. The chapter further examines the role and profile of the Data Scientist, by means of taking into consideration aspects such as functionality, academic background, and required skills. The authors inspect how Big Data is leading the world towards a new way of social construction, consumption, and processes.

Chapters 4 and 5 deal with the application of Big Data in the field of health care. Chapter 4 focuses on the use of data for in-patient care management in high-acuity spaces, such as operating rooms, intensive care units, and emergency departments. In addition, it discusses a variety of mathematical techniques to assist in managing and mitigating non-actionable alarm signals on monitored patients. Chapter 5 shows how the combination of novel biofeedback-based treatments producing large data sets with Big Data and Cloud-Dew technology can contribute to the greater good of patients with brain disorders. This approach is aimed at optimizing the therapy with regard to the current needs of the patient, improving the efficiency of the therapeutic process, and preventing patient from overstressing during the therapy. The preliminary results are documented using a case study confirming that the approach offers a viable way towards the greater good of the patients.

In the context of increased efforts dedicated to research on Big Data in agricultural and food research, Chap. 6 focuses on the presentation of an innovative and integrated Big Data e-infrastructure solution (AGINFRA+) that aims to enable the sharing of data, algorithms, and results in a scalable and efficient manner across different but interrelated research studies, with an application to the agriculture and food domain. The authors present three use cases for performing agri-food research with the help of the mentioned e-infrastructure. The chapter also analyses the new challenges and directions that will potentially arise for agriculture and food management and policing.

Chapter 7 aims to demonstrate the benefits of data collection and analysis to the maintenance and planning of current and future low-voltage networks. The authors review several agent-based modelling techniques and further present two case studies wherein these techniques are applied to energy modelling on a real low-voltage network in the UK. The chapter shows that Big Data Analytics of supply and demand can contribute to a more efficient usage of renewable sources, which will also result in cutting down carbon emissions.

It is common nowadays for customers to record their experiences in the form of online reviews and blogs. In Chap. 8, the authors examine the case of customer feedback at local, state, and national parks in the New York State Park System. The chapter discusses the design, development, and implementation of software systems that can download, organize, and analyse the voluminous text from the online reviews, analyse them using Natural Language Processing algorithms to perform sentiment analysis and topic modelling, and finally provide facility managers actionable insights to improve visitor experience.

Finally, Chap. 9 discusses the issue of data privacy, which has proven to be a challenge in the Big Data age, but which, nevertheless, can be addressed through modern cryptography. There are two types of solutions to tackle such problematic: one that makes data itself anonymous, but which degrades the value of the data, and the other one that uses Computation on Encrypted Data. This chapter introduces the latter and describes three prototype and pilot applications of the same within privacy-preserving statistics. The applications originate from R&D projects and collaborations between the Danish financial sector and Statistics Denmark.

The chapters contributed to this book should be of considerable interest and provide our readers with informative reading.

Birmingham, UK
Buckingham, UK
July 2018

Ali Emrouznejad
Vincent Charles

Contents

Chapter 1
Big Data for the Greater Good: An Introduction

Vincent Charles and Ali Emrouznejad

Abstract Big Data, perceived as one of the breakthrough technological developments of our times, has the potential to revolutionize essentially any area of knowledge and impact on any aspect of our life. Using advanced analytics techniques such as text analytics, machine learning, predictive analytics, data mining, statistics, and natural language processing, analysts, researchers, and business users can analyze previously inaccessible or unusable data to gain new insights resulting in better and faster decisions, and producing both economic and social value; it can have an impact on employment growth, productivity, the development of new products and services, traffic management, spread of viral outbreaks, and so on. But great opportunities also bring great challenges, such as the loss of individual privacy. In this chapter, we aim to provide an introduction into what Big Data is and an overview of the social value that can be extracted from it; to this aim, we explore some of the key literature on the subject. We also call attention to the potential 'dark' side of Big Data, but argue that more studies are needed to fully understand the downside of it. We conclude this chapter with some final reflections.

Keywords Big data · Analytics · Social value · Privacy

1.1 The Hype Around Big Data and Big Data Analytics

- Is it possible to predict whether a person will get some disease 24 h before any symptoms are visible?
- Is it possible to predict future virus hotspots?

V. Charles (✉)
Buckingham Business School, University of Buckingham, Buckingham, UK
e-mail: v.charles@buckingham.ac.uk

A. Emrouznejad
Aston Business School, Aston University, Birmingham, UK
e-mail: a.emrouznejad@aston.ac.uk

A. Emrouznejad and V. Charles (eds.), *Big Data for the Greater Good*,
Studies in Big Data 42, https://doi.org/10.1007/978-3-319-93061-9_1

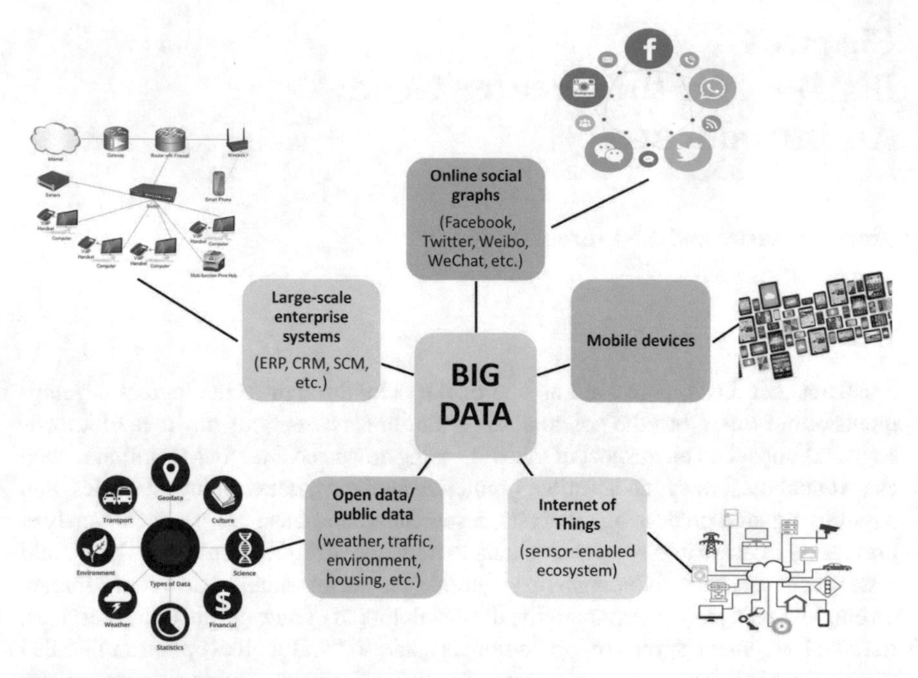

Fig. 1.1 Sources of big data. Adapted from Baesens et al. [2]

- Is it possible to predict when or where a fraud or crime will take place before it actually happens?
- It is possible to predict traffic congestion up to three hours in advance?
- Is it possible to predict terrorists' future moves?
- Can we support in a better way the wellbeing of people?

These are some of the questions that the Big Data age promises to have an answer for. We should note, at the outset, that some of these questions have already been answered, fostering new waves of creativity, innovation, and social change. But the true potential of Big Data in any of the areas mentioned is yet to be fully unlocked.

Big Data is a big phenomenon that for the past years has been fundamentally changing not only what we know, but also what we do, how we communicate and work, and how we cooperate and compete. It has an impact at the individual, organizational, and societal level, being perceived as a breakthrough technological development [18].

Today, we are witnessing an exponential increase in 'raw data', both human and machine-generated; human, borne from the continuous social interactions and doings among individuals, which led McAfee and Brynjolfsson [44] to refer to people as "walking data generators" (p. 5); machine-generated, borne from the continuous interaction among objects (generally coined the 'Internet-of-Things'), data which is generally collected via sensors and IP addresses. According to Baesens et al. [2], Big Data comes from five major sources (see Fig. 1.1 for a visual representation):

- *Large-scale enterprise systems*, such as enterprise resource planning, customer relationship management, supply chain management, and so on.
- *Online social graphs*, resulting from the interactions on social networks, such as Facebook, Twitter, Instagram, WeChat, and so on.
- *Mobile devices*, comprising handsets, mobile networks, and internet connection.
- *Internet-of-Things*, involving the connection among physical objects via sensors.
- *Open data/public data*, such as weather data, traffic data, environment and housing data, financial data, geodata, and so on.

Advanced analytics techniques such as text analytics, machine learning, predictive analytics, data mining, statistics, and natural language processing, are just as important as the Big Data itself. This may sound somehow obvious and trivial, but it is important to be clear about the weight that each holds in the discussion. On the one hand, were it not for the possibility to collect the large amounts of data being generated, the development of advanced analytics techniques would be irrelevant. On the other hand, the availability of the huge data would mean nothing without advanced analytics techniques to analyze it. Advanced analytics techniques and Big Data are, thus, intertwined. And thanks to Big Data Analytics, we now have the possibility to transform all of the data into meaningful information that can be explored at various levels of the organization or society [10].

It is useful to distinguish among three different types of analytics: descriptive, predictive, and prescriptive. It should be noted that today, nonetheless, most of the efforts are directed towards predictive analytics.

- *Descriptive analytics* help answer the question: *What happened?* It uses data aggregation and data mining (dashboards/scorecards and data visualization, among others). It helps to summarize and describe the past and is useful as it can shed light onto behaviors that can be further analyzed to understand how they might influence future results. For example, descriptive statistics can be used to show average pounds spent per household or total number of vehicles in inventory.
- *Predictive analytics* help answer the question; *What will or could happen?* It uses statistical models and forecasts techniques (regression analysis, machine learning, neural networks, golden path analysis, and so on). It helps to understand the future by means of providing estimates about the likelihood of a future outcome. For example, predictive analytics can be used to forecast customer purchasing patterns or customer behavior.
- *Prescriptive analytics* help answer the question: *How can we make it happen? Or what should we do?* It uses optimization and simulation algorithms (algorithms, machine learning, and computational modelling procedure, among others). It helps to advise on possible outcomes before decisions are made by means of quantifying the effect of future decisions. For example, prescriptive analytics can be used to optimize production or customer experience [16].

Despite the increased interest in exploring the benefits of Big Data emerging from performing descriptive, predictive, and/or prescriptive analytics, however, researchers and practitioners alike have not yet agreed on a unique definition of the concept of *Big Data* [5, 25, 36, 49]. We have performed a Google search of the most

Fig. 1.2 A broad view of the big data terminology

common terms associated with Big Data and Fig. 1.2 depicts these terms that we have compiled from across the big data spectrum.

The fact that there is no unique definition of Big Data is not necessarily bad, since this allows the possibility to explore facets of Big Data that may otherwise be constrained by a definition; but, at the same time, it is surprising if we consider that what was once considered to be a problem, that is, the collection, storage, and processing of the large amount of data, today is not an issue anymore; as a matter of fact, advanced analytics technologies are constantly being developed, updated, and used. In other words, we have the IT technology to support the finding of innovative insights from Big Data [14], while we lack a unified definition of Big Data. The truth is that indeed, this is not a problem. We may not all agree on *what* Big Data is, but we do all agree that Big Data *exists*. And *grows* exponentially. And this is sufficient because what this means is that we now have more information than ever before and we have the technology to perform analyses that could not be done before when we had smaller amounts of data.

Let us consider, for example, the case of WalMart, who has been using predictive technology since 2004. Generally, big retailers (WalMart included) would collect information about their transactions for the purpose of knowing how much they are selling. But WalMart took advantage of the benefits posed by Big Data, and took that extra step when it started analyzing the trillions of bytes' worth of sales data, looking for patterns and correlations. One of the things they were able to determine was what customers purchase the most ahead of a storm. And the answer was Pop-Tarts [68], a sugary pastry that requires no heating, lasts for an incredibly long period of time, and can be eaten at any meal. This insight allowed WalMart to optimize its supply of Pop-Tarts in the months or weeks leading up to a possible storm. The benefit for both the company and the customers is, thus, obvious.

Considering the above example, one of the conclusions we can immediately draw is that the main challenge in the Big Data age remains how to use the newly-generated data to produce the greatest value for organizations, and ultimately, for the society.

Einav and Levin [15], captured this view when they elegantly stated that Big Data's potential comes from the "identification of novel patterns in behaviour or activity, and the development of predictive models, that would have been hard or impossible with smaller samples, fewer variables, or more aggregation" (p. 2). We thus also agree with the position taken by Baesens et al. [2], who stated that "analytics goes beyond business intelligence, in that it is not simply more advanced reporting or visualization of existing data to gain better insights. Instead, analytics encompasses the notion of going beyond the surface of the data to link a set of explanatory variables to a business response or outcome" (p. 808).

It is becoming increasingly clear that Big Data is creating the potential for significant innovation in many sectors of the economy, such as science, education, healthcare, public safety and security, retailing and manufacturing, e-commerce, and government services, just to mention a few—we will discuss some of this potential later in the chapter. For example, according to a 2013 Report published by McKinsey Global Institute, Big Data analytics is expected to generate up to $190 billion annually in healthcare cost savings alone by 2020 [45]. Concomitantly, it is also true that despite the growth of the field, Big Data Analytics is still in its incipient stage and comprehensive predictive models that tie together knowledge, human judgement and interpretation, commitment, common sense and ethical values, are yet to be developed. And this is one of the main challenges and opportunities of our times. The true potential of Big Data is yet to be discovered.

We conclude this section with the words of Watson [66], who stated that "The keys to success with big data analytics include a clear business need, strong committed sponsorship, alignment between the business and IT strategies, a fact-based decision-making culture, a strong data infrastructure, the right analytical tools, and people skilled in the use of analytics" (p. 1247).

1.2 What Is Big Data?

The term 'Big Data' was initially used in 1997 by Michael Cox and David Ellsworth, to explain both the data visualization and the challenges it posed for computer systems [13]. To say that Big Data is a new thing is to some extent erroneous. Data have always been with us; it is true, however, that during the 1990s and the beginning of the 2000s, we experienced an increase in IT-related infrastructure, which allowed to store the data that was being produced. But most of the time, these data were simply that: stored—and most probably forgotten. Little value was actually being extracted from the data. Today, besides the required IT technology, our ability to generate data has increased dramatically—as mentioned previously, we have more information than ever before, but what has really changed is that we can now analyze and interpret the data in ways that could not be done before when we had smaller amounts of data. And this means that Big Data has the potential to revolutionize essentially any area of knowledge and any aspect of our life.

Big Data has received many definitions and interpretations over time and a unique definition has not been yet reached, as indicated previously. Hammond [25] associated Big Data with evidence-based decision-making; Beyer and Laney [5] defined it as high volume, high velocity, and/or high variety information assets; and Ohlhorst [49] described Big Data as vast data sets which are difficult to analyze or visualize with conventional information technologies. Today, it is customary to define Big Data in terms of data characteristics or dimensions, often with names starting with the letter 'V'. The following four dimensions are among the most often encountered [37]:

Volume: It refers to the large amount of data created every day globally, which includes both simple and complex analytics and which poses the challenge of not just storing it, but also analyzing it [37]. It has been reported that 90% of the existent data has been generated in the past two years alone [30]. It is also advanced that by 2020, the volume of data will be 40 ZB, 300 times bigger than the volume of data in 2005 [27].

Velocity: It refers to the speed at which new data is generated as compared to the time window needed to translate it into intelligent decisions [37]. It is without doubt that in some cases, the speed of data creation is more important than the volume of the data; IBM [29] considered this aspect when they stated that "for time-sensitive processes such as catching fraud, big data must be used as it streams into your enterprise in order to maximize its value". Real-time processing is also essential for businesses looking to obtain a competitive advantage over their competitors, for example, the possibility to estimate the retailers' sales on a critical day of the year, such as Christmas [44].

Variety: It encapsulates the increasingly different types of data, structured, semi-structured, and unstructured, from diverse data sources (e.g., web, video and audio data, sensor data, financial data and transactional applications, log files and click streams, GPS signals from cell phones, social media feeds, and so on), and in different sizes from terabytes to zettabytes [30, 37]. One of the biggest challenges is posed by unstructured data. Unstructured data is a fundamental concept in Big Data and it refers to data that has no rules attached to it, such as a picture or a voice recording. The challenge is how to use advanced analytics to make sense of it.

Veracity: It refers to the trustworthiness of the data. In its 2012 Report, IBM showed that 1 in 3 business leaders don't trust the information they use to make decisions [28]. One of the reasons for such phenomenon is that there are inherent discrepancies in the data, most of which emerge from the existence of unstructured data. This is even more interesting if we consider that, today, most of the data is unstructured. Another reason is the presence of inaccuracies. Inaccuracies can be due to the data being intrinsically inaccurate or from the data becoming inaccurate through processing errors [37].

Building upon the 4 Vs, Baesens et al. [2], stated that:

> The 4 V definition is but a starting point that outlines the perimeters. The definition does not help us to determine what to do inside the perimeters, how to innovatively investigate and analyze big data to enhance decision making quality, how to anticipate and leverage the transformational impacts of big data, or how best to consider scope as well as scale impacts of big data (p.807).

As such, they argued for the necessity to include a 5th V, namely *Value*, to complement the 4 V framework. It should be noted that, by some accounts, there are as many as 10 Vs [40].

Charles and Gherman [9] argued that the term Big Data is a misnomer, stating that while the term in itself refers to the large volume of data, Big Data is essentially about the phenomenon that we are trying to record and the hidden patterns and complexities of the data that we attempt to unpack. With this view, the authors advanced an expanded model of Big Data, wherein they included three additional dimensions, namely the 3 Cs: Context, Connectedness, and Complexity. The authors stated that understanding the Context is essential when dealing with Big Data, because "raw data could mean anything without a thorough understanding of the context that explains it" (p. 1072); Connectedness was defined as the ability to understand Big Data in its wider Context and within its ethical implications; and Complexity was defined from the perspective of having the skills to survive and thrive in the face of complex data, by means of being able to identify the key data and differentiate the information that truly has an impact on the organization.

Figure 1.3 briefly summarizes the discussed characteristics of Big Data. Presenting them all goes beyond the scope of the present chapter, but we hope to have provided a flavor of the various dimensions of Big Data. Having, thus, highlighted these, along with the existing debate surrounding the very definition of Big Data, we will now move towards presenting an overview of the social value that Big Data can offer.

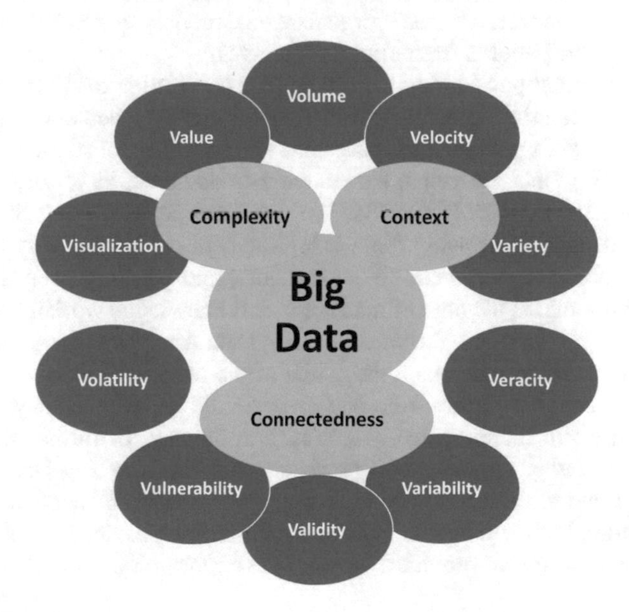

Fig. 1.3 Big data characteristics

1.3 The Social Value of Big Data

Value creation in a Big Data perspective includes both the traditional economic dimension of value and the social dimension of value [54]. Today, we are yet to fully understand how organizations actually translate the potential of Big Data into the said value [24]. Generally, however, stories of Big Data's successes have tended to come from the private sector, and less is known about its impact on social organizations. Big Data can, nonetheless, drive big social change, in fields such as education [8, 50], healthcare [52], and public safety and security [48], just to mention a few.

Furthermore, the social value can be materialized as employment growth [17], increased productivity [67], increased consumer surplus [7], new products and services, new markets and better marketing [11], and so on. Governments, for instance, can use big data to, "enhance transparency, increase citizen engagement in public affairs, prevent fraud and crime, improve national security, and support the wellbeing of people through better education and healthcare" (p. 81) [36].

Loebbecke and Picot [38] published a viewpoint focusing on the impact of 'datification' [21] on business models and employment and the consequential societal transformation. More recently, Günther et al. [24] have conducted an in-depth systematic review of Information Systems literature on the topic and identified two socio-technical features of Big Data that influence value realization: portability and interconnectivity. The authors further argue that, in practice, "organizations need to continuously realign work practices, organizational models, and stakeholder interests in order to reap the benefits from big data" (p. 191).

In 2015, drawing upon Markus and Silver [42] and Kallinikos [32], Markus [40], elegantly drew attention to an important societal impact when she stated that the "source of Big Data's potentially transformative power lies not so much in the *characteristics* of Big Data, on which […] authors concentrate, as it does in what Big Data's characteristics *afford*" (p. 58). She continues by clarifying the statement: "Together with today's growing 'data lakes,' algorithms increasingly afford organizations the ability to automate the operational, and possibly even the strategic, decision making that is the core of managers' and knowledge workers' jobs".

As previously mentioned, the value that Big Data Analytics can unleash is great, but we are yet to fully understand the extent of the benefits. Empirical studies that consider the creation of value from Big Data Analytics are nowadays growing in number, but are still rather scarce. We, thus, join the call for further studies in the area. In the following lines, we will proceed to explore how Big Data can inform social change and to this aim, we present some of the advancements made in three different sectors. It should be noted that the information provided is not exhaustive, as our main intention is to provide a flavor of the opportunities brought about by the Big Data age.

1.3.1 Healthcare

Is it possible to predict whether a person will get some disease 24 hours before any symptoms appear? It is generally considered that the healthcare system is one of the sectors that will benefit the most from the existence of Big Data Analytics. Let us explore this in the following lines.

There is certain consensus that some of the challenges faced by the healthcare sector include the inadequate integration of the healthcare systems and the poor healthcare information management [6]. The healthcare sector in general amasses a large amount of information, which nonetheless, results today in unnecessary increases in medical costs and time for both healthcare service providers and patients. Researchers and hospital managers alike are thus interested in how this information could be used instead to deliver a high-quality patient experience, while also improving organizational and financial performance and meeting future market needs [1, 23, 35, 65].

Watson [66] and Raghupathi and Raghupathi [52] advanced that Big Data Analytics can support evidence-based decision-making and action taking in healthcare. In this sense, Cortada et al. [12] performed a study and found that only 42% of the healthcare organizations surveyed supported their decision-making process with Big Data Analytics and only 16% actually had the necessary skills and experience to use Big Data Analytics. The value that can be generated in general goes, thus, far beyond the one that is created today.

But beyond improving profits and cutting down on operating costs, Big Data can help in other ways, such as curing disease or detecting and predicting epidemics. Big Data Analytics can help to collect and analyze the health data that is constantly being generated for faster responses to individual health problems; ultimately, for the betterment of the patient. It is now a well-known fact that with the help of Big Data Analytics, real-time datasets have been collected, modelled, and analyzed and this has helped speed up the development of new flu vaccines, identifying and containing the spread of viral outbreaks such as the Dengue fever or even Ebola.

Furthermore, we can only imagine for now what we would be able to do if, for example, we would collect all the data that is being created during every single doctor appointment. There are a variety of activities that happen during the routine medical examinations, which are not necessarily recorded, especially if the results turn out to be within some set parameters (in other words, if the patient turns out to be healthy): the doctor will take the body temperature and blood pressure, look into the eyes to see the retina lining, use an otoscope to look into the ears, listen to heartbeat for regularity, listen to breathing in the lungs, and so on—all these data could help understand as much about a patient as possible, and as early in his or her life as possible. This, in turn, could help identify warning signs of illness with time in advance, preventing further advancement of the disease, increasing the odds of success of the treatment, and ultimately, reducing the associated expenses. Now, to some extent, collecting this kind of granular data about an individual is possible due to smart phones, dedicated wearable devices, and specialized apps, which can collect data, for example, on how many steps a day a person walks and on the number of daily calories consumed, among others. But the higher value that these datasets hold is yet to be explored.

Psychiatry is a particular branch of medicine that could further benefit from the use of Big Data Analytics and research studies to address such matter have just recently started to emerge. It is well known that in psychiatric treatments, there are treatments that are proven to be successful, but what cannot be predicted generally is who they are going to work for; we cannot yet predict a patient's response to a specific treatment. What this means, in practical terms, is that most of the times a patient would have to go through various trials with various medicines, before identifying that works the best for the patient in question. In a recent contribution, Gillan and Whelan [22] emphasized the importance of Big Data and robust statistical methodologies in treatment prediction research, and in so doing, they advocated for the use of machine-learning approaches beyond exploratory studies and toward model validation. The practical implications of such endeavors are rather obvious.

According to Wang et al. [64], the healthcare industry in general has not yet fully understood the potential benefits that could be gained from Big Data Analytics. The authors further stated that most of the potential value creation is still in its infancy, as predictive modelling and simulation techniques for analyzing healthcare data as a whole have not been yet developed. Today, one of the biggest challenges for healthcare organizations is represented by the missing support infrastructure needed for translating analytics-derived knowledge into action plans, a fact that is particularly true in the case of developing countries.

1.3.2 Agriculture

It is to some extent gratuitous to say that, in the end, nothing is more important than our food supply. Considering that we still live in a world in which there are people dying of starvation, it comes as quite a surprise to note that about a third of the food produced for human consumption is lost of wasted every year [39]. The agriculture sector is thus, in desperate need of solutions to tackle problems such as inefficiencies in planting, harvesting, and water use and trucking, among others [57]. The Big Data age promises to help. For example, Big Data Analytics can help farmers simulate the impact of water, fertilizer, and pesticide, and engineer plants that will grow in harsh climatic conditions; it can help to reduce waste, increase and optimize production, speed up plant-growth, and minimize the use of scarce resources, such as water.

Generally speaking, Big Data Analytics has not yet been widely applied in agriculture [33]. Nonetheless, there is increasing evidence of the use of digital technologies [4, 26]; and bio-technologies [53] to support agricultural practices. This is termed as *smart farming* [61], a concept that is closely related to *sustainable agriculture* [55].

Farmers have now started using high-technology devices to generate, record, and analyze data about soil and water conditions and weather forecast in order to extract insights that would assist them in refining their decision-making process. Some examples of tools being used in this regard include: agricultural drones (for fertilizing crops), satellites (for detecting changes in the field); and sensors on field (for collecting information about weather conditions, soil moisture and humidity, and so on).

As of now, Big Data Analytics in agriculture has resulted in a number of research studies in several areas—we herewith mention some of the most recent ones: crops [63], land [3], remote sensing [47], weather and climate change [58], animals' research [34], and food availability and security [19, 31].

According to Wolfert et al. [69], the applicability of Big Data in agriculture faces a series of challenges, among which: data ownership and security and privacy issues, data quality, intelligent processing and analytics, sustainable integration of Big Data sources, and openness of platforms to speed up innovation and solution development. These challenges would need, thus, to be addressed in order to expand the scope of Big Data applications in agriculture and smart farming.

1.3.3 Transportation

Traffic congestion and parking unavailability are few examples of major sources of traffic inefficiency. Worldwide. But how about if we could change all that? How about if we could predict traffic jams hours before actually taking place and use such information to reach our destinations within lesser time? How about if we could be able to immediately find an available parking space and avoid frustration considerably? Transportation is another sector that can greatly benefit from Big Data. There is a huge amount of data that is being created, for example, from the sat nav installed in vehicles, as well as the embedded sensors in infrastructure.

But what has been achieved so far with Big Data Analytics in transportation? One example is the development of the ParkNet system ("ParkNet at Rutgers", n/a), a wireless sensing network developed in 2010 which detects and provides information regarding open parking spaces. The way it works is that a small sensor is attached to the car and an on-board computer collects the data which is uploaded to a central server and then processed to obtain the parking availability.

Another example is VTrack [60], a system for travel time estimation using sensor data collected by mobile phones that addresses two key challenges: reducing energy consumption and obtaining accurate travel time estimates. In the words of the authors themselves:

> Real-time traffic information, either in the form of travel times or vehicle flow densities, can be used to alleviate congestion in a variety of ways: for example, by informing drivers of roads or intersections with large travel times ("hotspots"); by using travel time estimates in traffic-aware routing algorithms to find better paths with smaller expected time or smaller variance; by combining historic and real-time information to predict travel times in specific areas at particular times of day; by observing times on segments to improve operations (e.g., traffic light cycle control), plan infrastructure improvements, assess congestion pricing and tolling schemes, and so on.

A third example is VibN [46], a mobile sensing application capable of exploiting multiple sensor feeds to explore live points of interest of the drivers. Not only that, but it can also automatically determine a driver's personal points of interest.

Lastly, another example is the use of sensors embedded in the car that could be able to predict when the car would break down. A change in the sound being emitted by the engine or a change in the heat generated by certain parts of the car—all these data and much more could be used to predict the increased possibility of a car to break down and allow the driver to take the car to a mechanic prior to the car actually breaking down. And this is something that is possible with Big Data and Big Data Analytics and associated technologies.

To sum up, the Big Data age presents opportunities to use traffic data to not only solve a variety of existent problems, such as traffic congestion and equipment fault, but also predict traffic congestion and equipment fault before it actually happens. Big Data Analytics can, thus, be used for better route planning, traffic monitoring and management, and logistics, among others.

1.4 The Good… but What About the Bad?

As early as 1983 and as recent as 2014, Pool [51] and Markus [41], respectively, warned that Big Data is not all good. Any given technology is argued to have a *dual* nature, bringing both positive and negative effects that we should be aware of. Below we briefly present two of the latter effects.

In the context of Big Data and advanced analytics, a negative aspect, which also represents one of the most sensitive and worrisome issues, is the privacy of personal information. When security is breached, privacy may be compromised and loss of privacy can in turn result in other harms, such as identity theft and cyberbullying or cyberstalking. "[…] There is a great public fear about the inappropriate use of personal data, particularly through the linking of data from multiple sources. Managing privacy is effectively both a technical and a sociological problem, and it must be addressed jointly from both perspectives to realize the promise of big data" (p. 122) [49]. Charles et al. [10] advocated that, in the age of Big Data, there is a necessity to create new principles and regulations to cover the area of privacy of information, although who exactly should create these new principles and regulations is a rather sensitive question.

Marcus [40] highlights another danger, which she calls 'displacement by automation'. She cites Frey and Osborne [20] and notes that "Oxford economists […] computed (using machine learning techniques) that 47% of US employment is at risk to automation, though mainly at the lower end of the occupational knowledge and skill spectrum" (p. 58). We may wonder: *Is this good or is this bad?* On the one hand, behavioral economists would argue that humans are biased decision-makers, which would support the idea of automation. But on the other hand, what happens when individuals gain the skills necessary to use automation, but know very little about the underlying assumptions and knowledge domain that make automation possible?

It is without much doubt that The Bad or *dark* side of the Big Data age cannot be ignored and should not be treated with less importance than it merits, but more in-depth research is needed to explore and gain a full understanding of its negative implications and how these could be prevented, diminished, or corrected.

1.5 Conclusion

In this introductory chapter, we have aimed to provide an overview of the various dimensions and aspects of Big Data, while also exploring some of the key research studies that have been written and related applications that have been developed, with a special interest on the societal value generated.

It has been advocated that advanced analytics techniques should support, but not replace, human decision-making, common sense, and judgment (Silver [56]; [9]. We align with such assessment that indeed, without the qualities mentioned above, Big Data is, most likely, meaningless. But we also need to be pragmatic and accept evidence that points to the contrary. Markus [40], for example, pointed out that today there is evidence of automated decision-making, with minimal human intervention:

> By the early 2000s, nearly 100% of all home mortgage underwriting decisions in the United States were automatically made [43]. Today, over 60% of all securities trading is automated, and experts predict that within ten years even the writing of trading algorithms will be automated, because humans will not be fast enough to anticipate and react to arbitrage opportunities [59]. IBM is aggressively developing Watson, its Jeopardy game show-winning software, to diagnose and suggest treatment options for various diseases [62]. In these and other developments, algorithms and Big Data [...] are tightly intertwined (p. 58).

It is, thus, not too bold to say that the potential applications and social implications brought by the Big Data age are far from being entirely understood and continuously change, even as we speak. Without becoming too philosophical about it, we would simply like to conclude the above by saying that a Big Data Age seems to require a Big Data Mind; and this is one of the greatest skills that a Big Data Scientist could possess.

Today, there are still many debates surrounding Big Data and one of the most prolific ones involves the questioning of the very existence of Big Data, with arguments in favor or against Big Data. But this is more than counter-productive. Big Data is here to say. In a way, the Big Data age can be compared to the transition from the Stone Age to the Iron Age: it is simply the next step in the evolution of the human civilization and it is, quite frankly, irreversible. And just because we argue over its presence does not mean it will disappear. The best we can do is accept its existence as the natural course of affairs and instead concentrate all our efforts to chisel its path forward in such a way so as to serve a Greater Good.

We conclude by re-stating that although much has been achieved until the present time, the challenge remains the insightful interpretation of the data and the usage of the knowledge obtained for the purposes of generating the most economic and social value [10]. We join the observation made by other researchers according to which many more research studies are needed to fully understand and fully unlock the societal value of Big Data. Until then, we will most likely continue to live in a world wherein individuals and organizations alike collect massive amounts of data with a 'just in case we need it' approach, trusting that one day, not too far away, we will come to crack the *Big Data Code*.

References

1. R. Agarwal, G. Gao, C. DesRoches, A.K. Jha, Research commentary—the digital transformation of healthcare: current status and the road ahead. Inf. Syst. Res. **21**(4), 796–809 (2010)
2. B. Baesens, R. Bapna, J.R. Marsden, J. Vanthienen, J.L. Zhao, Transformational issues of big data and analytics in networked business. MIS Q. **40**(4), 807–818 (2016)
3. B. Barrett, I. Nitze, S. Green, F. Cawkwell, Assessment of multi-temporal, multisensory radar and ancillary spatial data for grasslands monitoring in Ireland using machine learning approaches. Remote Sens. Environ. **152**(2), 109–124 (2014)
4. W. Bastiaanssen, D. Molden, I. Makin, Remote sensing for irrigated agriculture: examples from research and possible applications. Agric. Water Manage. **46**(2), 137–155 (2000)
5. M.A. Beyer, D. Laney, The Importance of 'Big Data': A Definition, META Group (now Gartner) [online] (2012) https://www.gartner.com/doc/2057415/importance-big-data-definition. Accessed 10 Aug 2017
6. T. Bodenheimer, High and rising health care costs. Part 1: seeking an explanation. Ann. Intern. Med. **142**(10), 847–854 (2005)
7. E. Brynjolfsson, A. Saunders, *Wired for Innovation* (The MIT Press, Cambridge, MA, How Information Technology is Reshaping the Economy, 2010)
8. T.G. Cech, T.K. Spaulding, J.A. Cazier, in *Proceedings of the Twenty-First Americas Conference on Information Systems*. Applying business analytic methods to improve organizational performance in the public school system, Puerto Rico, 13–15 Aug (2015)
9. V. Charles, T. Gherman, Achieving competitive advantage through big data. Strategic implications. Middle-East. J. Sci. Res. **16**(8), 1069–1074 (2013)
10. V. Charles, M. Tavana, T. Gherman, The right to be forgotten—is privacy sold out in the big data age? Int. J. Soc. Syst. Sci. **7**(4), 283–298 (2015)
11. I.D. Constantiou, J. Kallinikos, New games, new rules: big data and the changing context of strategy. J. Inf. Technol. **30**(1), 44–57 (2015)
12. J.W. Cortada, D. Gordon, B. Lenihan, *The Value of Analytics in Healthcare: From Insights to Outcomes* (IBM Global Business Services, Somers, NY, 2012)
13. M. Cox, D. Ellsworth, in *Proceedings of the 8th IEEE Conference on Visualization*. Application-controlled demand paging for out-of-core visualization (IEEE Computer Society Press, Los Alamitos, CA, 1997)
14. T.H. Davenport, P. Barth, R. Bean, How 'big data' is different. MIT Sloan Manage. Rev. **54**(1), 43–46 (2012)
15. L. Einav, J.D. Levin, 'The Data Revolution and Economic Analysis', Prepared for NBER Innovation Policy and the Economy Conference [online] April (2013) http://www.nber.org/papers/w19035.pdf. Accessed 30 June 2018
16. A. Emrouznejad, Big Data Optimization: Recent Developments and Challenges. In the series of "Studies in Big Data", Springer. ISBN: 978-3-319-30263-8 (2016)
17. J. Enck, T. Reynolds *Network Developments in Support of Innovation and User Needs*, No. 164, (OECD Publishing, 2009)
18. R.G. Fichman, B.L. Dos Santos, Z. Zheng, Digital innovation as a fundamental and powerful concept in the information systems curriculum. MIS Q. **38**(2), 329–353 (2014)

19. R. Frelat et al., Drivers of household food availability in sub-Saharan Africa based on big data from small farms. Proc. Natl. Acad. Sci. U.S. Am. **113**(2), 458–463 (2016)
20. C.B. Frey, M.A. Osborne, *The Future of Employment: How susceptible are jobs to computerization?* (Oxford Martin Programme on the Impacts of Future Technology, Oxford, 2013)
21. R. Galliers, S. Newell, G. Shanks, H. Topi, Call for papers for the special issue: the challenges and opportunities of 'datification'; Strategic impacts of 'big' (and 'small') and real time data—for society and for organizational decision makers. J. Strateg. Inf. Syst. **24**, II–III (2015)
22. C.M. Gillan, R. Whelan, What big data can do for treatment in psychiatry. Curr. Opin. Behav. Sci. **18**, 34–42 (2017)
23. J.M. Goh, G. Gao, R. Agarwal, Evolving work routines: adaptive routinization of in-formation technology in healthcare. Inf. Syst. Res. **22**(3), 565–585 (2011)
24. W.A. Günther, M.H. Rezazade Mehrizi, M. Huysman, F. Feldberg, Debating big data: a literature review on realizing value from big data. J. Strateg. Inf. Syst. **26**, 191–209 (2017)
25. K.J. Hammond, 'The value of big data isn't the data', Harvard Business Review, May [online] (2013) http://blogs.hbr.org/cs/2013/05/the_value_of_big_data_isnt_the.html. Accessed 13 July 2017
26. I. Hashem et al., The rise of "big data" on cloud computing: review and open research issues. Inf. Syst. **47**, 98–115 (2015)
27. R. Herschel, V.M. Miori, Ethics and big data. Technol. Soc. **49**, 31–36 (2017)
28. IBM, The Four V's of Big Data. [online] http://www.ibmbigdatahub.com/infographic/four-vs-big-data. Accessed 30 June 2018
29. Technology Advice, The Four V's of Big Data [online] (2013) https://technologyadvice.com/blog/information-technology/the-four-vs-of-big-data/. Accessed 20 July 2017
30. IBM, What is Big Data? [online] (2016) https://www.ibm.com/analytics/hadoop/big-data-analytics. Accessed 20 Nov 2017
31. Á. Jóźwiaka, M. Milkovics, Z. Lakne, A network-science support system for food chain safety: a case from Hungarian cattle production. Int. Food Agribusiness Manage. Rev. Special Issue, **19**(A) (2016)
32. J. Kallinikos, Governing Through Technology: Information Artefacts and Social Practice. (Palgrave Macmillan, Basingstoke, UK, 2011)
33. A. Kamilaris, A. Kartakoullis, F.X. Prenafeta-Boldu, A review on the practice of big data analysis in agriculture. Comput. Electron. Agric. **143**, 23–37 (2017)
34. C. Kempenaar et al., *Big Data Analysis for Smart Farming*, vol. 655 (Wageningen University & Research, s.l., 2016)
35. J.I. Ker, Y. Wang, M.N. Hajli, J. Song, C.W. Ker, Deploying lean in healthcare: evaluating information technology effectiveness in US hospital pharmacies. Int. J. Inf. Manage. **34**(4), 556–560 (2014)
36. G.-H. Kim, S. Trimi, J.-H. Chung, Big data applications in the government sector. Commun. ACM. **57**(3), 78–85 (2014).
37. D. Laney, 3D Data Management: controlling data volume, velocity and variety. Applications delivery strategies, META Group (now Gartner) [online] (2001) http://blogs.gartner.com/doug-laney/files/2012/01/ad949-3D-Data-Management-Controlling-Data-Volume-Velocity-and-Variety.pdf. Accessed 1 Aug 2017
38. C. Loebbecke, A. Picot, Reflections on societal and business model transformation arising from digitization and big data analytics: a research agenda. J. Strateg. Inf. Syst. **24**(3), 149–157 (2015). https://doi.org/10.1016/j.jsis.2015.08.002
39. C. Magnin, How big data will revolutionize the global food chain [online] (McKinsey & Company, 2016). https://www.mckinsey.com/business-functions/digital-mckinsey/our-insights/how-big-data-will-revolutionize-the-global-food-chain. Accessed 13 December 2017
40. L. Markus, New games, new rules, new scoreboards: the potential consequences of big data. J. Inf. Technol. **30**(1), 58–59 (2015)
41. M.L. Markus, Information Technology and Organizational Structure, in *Information Systems and Information Technology, Computing Handbook*, ed. by H. Topi, A. Tucker, vol. Ii. (Chapman and Hall, CRC Press, 2014), p. 67, 61–22

42. M.L. Markus, M.S. Silver, A foundation for the study of It effects: a new look at desanctis and poole's concepts of structural features and spirit. Journal of the AIS, **9**(10/11), 609–632 (2008).
43. M.L. Markus, A. Dutta, C.W. Steinfield, R.T. Wigand, The Computerization Movement in the Us Home Mortgage Industry: Automated underwriting from 1980 to 2004, in *Computerization Movements and Technology Diffusion: From mainframes to ubiquitous computing*, ed. by K.L. Kraemer, M.S. Elliott (Information Today, Medford, NY, 2008), pp. 115–144
44. A. McAfee, E. Brynjolfsson, Big data: the management revolution. Harvard Bus. Rev. **90**(10), 60–68 (2012)
45. McKinsey Global Institute, Game changers: five opportunities for US growth and renewal, [online] July (2013), http://www.mckinsey.com/insights/americas/us_game_changers. Accessed 13 Dec 2017
46. E. Miluzzo, M. Papandrea, N.D. Lane, A.M. Sarroff, S. Giordano, A.T. Campbell, In *Proceedings of 1st International Symposium on from Digital Footprints to Social and Community Intelligence*. Tapping into the vibe of the city using vibn, a continuous sensing application for smartphones (Beijing, China: ACM, 2011), pp. 13–18
47. S. Nativi et al., Big data challenges in building the global earth observation system of systems. Environ. Model Softw. **68**(1), 1–26 (2015)
48. S. Newell, M. Marabelli, Strategic opportunities (and challenges) of algorithmic decision-making: a call for action on the long-term societal effects of 'datafication'. J. Strateg. Inf. Syst. **24**(1), 3–14 (2015). https://doi.org/10.1016/j.jsis.2015.02.001
49. F. Ohlhorst, *Big Data Analytics: Turning Big Data into Big Money* (Wiley, Hoboken, NJ, 2013)
50. Z.A. Pardos, Big data in education and the models that love them. Curr. Opin. Behav. Sci. **18**, 107–113 (2017)
51. I.d.S. Pool, Forecasting the Telephone: A Retrospective Technology Assessment of the Telephone (Ablex, Norwood, NJ, 1983)
52. W. Raghupathi, V. Raghupathi, Big data analytics in healthcare: promise and potential. Health Inf. Sci. Syst. **2**(3), 1–10 (2014). https://doi.org/10.1186/2047-2501-2-3
53. H. Rahman, D. Sudheer Pamidimarri, R. Valarmathi, M. Raveendran, *Omics: Applications in Biomedical* (CRC PressI Llc, Agriculture and Environmental Sciences, s.l, 2013)
54. G. Secundo, P. Del Vecchio, J. Dumay, G. Passiante, Intellectual capital in the age of big data: establishing a research agenda. J. Intellect. Capital **18**(2), 242–261 (2017)
55. R. Senanayake, Sustainable agriculture: definitions and parameters for measurement. J. Sustain. Agric. **1**(4), 7–28 (1991)
56. M.S. Silver, Systems That Support Decision Makers: Description and analysis (John Wiley & Sons, Chichester, UK, 1991)
57. T. Sparapani, How Big Data and Tech Will Improve Agriculture, from Farm to Table. [online] (Forbes, 2017). https://www.forbes.com/sites/timsparapani/2017/03/23/how-big-data-and-tech-will-improve-agriculture-from-farm-to-table/#503f16c25989. Accessed 13 December 2017
58. K. Tesfaye et al., Targeting drought-tolerant maize varieties in southern Africa: a geospatial crop modeling approach using big data. Int Food Agribusiness Manage. Rev. **19**(A), 1–18 (2016)
59. The Government Office for Science, *Foresight: The Future of Computer Trading in Financial Markets* (Final Project Report, London, 2010)
60. A. Thiagarajan, L. Ravindranath, K. LaCurts, S. Madden, H. Balakrishnan, S. Toledo, J. Eriksson, in *Proceedings of the 7th ACM Conference on Embedded Networked Sensor Systems*. Vtrack: accurate, energy-aware road traffic delay estimation using mobile phones (ACM, Berkeley, California, 2009), pp. 85–98
61. A.C. Tyagi, Towards a second green revolution. Irrig. Drainage **65**(4), 388–389 (2016)
62. N. Ungerleider, IBM's Watson is ready to see you now—in your dermatologist's office. Fast Company [online] May (2014) http://www.fastcompany.com/3030723/ibms-watson-is-ready-to-see-you-now-in-yourdermatologists-office. Accessed 10 January 2018

63. G. Waldhoff, C. Curdt, D. Hoffmeister, G. Bareth, Analysis of multitemporal and multisensor remote sensing data for crop rotation mapping. Int. Arch. Photogrammetry Remote Sensing Spat. Inf. Sci. **25**(1), 177–182 (2012)
64. Y. Wang, L. Kung, T.A. Byrd, Big data analytics: understanding its capabilities and potential benefits for healthcare organizations. Technol. Forecast. Soc. Change **126**, 3–13 (2018)
65. Y. Wang, L. Kung, C. Ting, T.A. Byrd, in *2015 48th Hawaii International Conference*. Beyond a technical perspective: understanding big data capabilities in health care. System Sciences (HICSS) (IEEE, 2015), pp. 3044–3053
66. H.J. Watson, Tutorial: big data analytics: concepts, technologies, and applications. Commun. Assoc. Inf. Syst. **34**(1), 1247–1268 (2014)
67. P. Weill, S. Woerner, Thriving in an increasingly digital ecosystem. MIT Sloan Manage. Rev. **56**(4), 27–34 (2015)
68. Why Walmart Always Stocks Up On Strawberry Pop-Tarts Before a Hurricane (2017). [online] August (2014) http://www.countryliving.com/food-drinks/a44550/walmart-strawberry-pop-tarts-before-hurricane/. Accessed 10 January 2018
69. S. Wolfert, L. Ge, C. Verdouw, M.J. Bogaardt, Big data in smart farming—a review. Agric. Syst. **153**, 69–80 (2017)

Prof. V. Charles is the Director of Research and Professor of Management Science at Buckingham Business School, The University of Buckingham, UK. He is an Honorary Professor at MSM, The Netherlands; an Affiliated Distinguished Research Professor at CENTRUM Católica GBS, Peru; a Fellow of the Pan-Pacific Business Association, USA; a Council Member of the Global Urban Competitiveness Project, USA; an HETL Liaison to Peru, HETL USA; and a WiDS Ambassador, WiDS Buckingham (in collaboration with Stanford University, USA). He is a PDRF from NUS, Singapore and a certified Six-Sigma MBB. He holds executive certificates from the HBS, MIT, and IE Business School. He has 20 years of teaching, research, and consultancy experience in various countries, in the fields of applied quantitative analytics (big data). He has over 120 publications. He holds international honours/awards, among which the 2016 Cum Laude Award from the Peruvian Marketing Society; the most innovative study to Peru's Think-Tank of the Year 2013; the highly commended award of Emerald/CLADEA and the intellectual contribution award of CENTRUM for three and seven consecutive years, respectively. He has developed the 4^3 Architecture for Service Innovation. His area of focus includes productivity, quality, efficiency, effectiveness, competitiveness, innovation, and design thinking.

Ali Emrouznejad is a Professor and Chair in Business Analytics at Aston Business School, UK. His areas of research interest include performance measurement and management, efficiency and productivity analysis as well as data mining and big data. Dr Emrouznejad is Editor of (1) *Annals of Operations Research*, (2) Associate Editor of *RAIOR-Operations Research*, (3) Associate Editor of *Socio-Economic Planning Sciences*, (4) Associate Editor of *IMA journal of Management Mathematics*, and (5) Senior Editor of *Data Envelopment Analysis journal*. He is also Guest Editor to several journals including *European Journal of Operational Research, Journal of Operational Research Society, Journal of Medical Systems* and *International Journal of Energy Management Sector*. He is also member of editorial boards other scientific journals. He has published over 120 articles in top ranked journals, he is also author / editor of several books including (1) "*Applied Operational Research with SAS*" (CRC Taylor & Francis), (2) "*Big Data Optimization*" (Springer), (3)

"P*erformance Measurement with Fuzzy Data Envelopment Analysis*" (Springer), (4) "*Managing Service Productivity*" (Springer), (5) "*Fuzzy Analytics Hierarchy Process*" (CRC Taylor & Francis), and (6) "*Handbook of Research on Strategic Performance Management and Measurement*" (IGI Global). For further details please visit http://www.emrouznejad.com/.

Chapter 2
Big Data Analytics and Ethnography: Together for the Greater Good

Vincent Charles and Tatiana Gherman

Abstract *Ethnography* is generally positioned as an approach that provides deep insights into human behaviour, producing 'thick data' from small datasets, whereas *big data analytics* is considered to be an approach that offers 'broad accounts' based on large datasets. Although perceived as antagonistic, ethnography and big data analytics have in many ways, a shared purpose; in this sense, this chapter explores the intersection of the two approaches to analysing data, with the aim of highlighting both their similarities and complementary nature. Ultimately, this chapter advances that ethnography and big data analytics can work together to provide a more comprehensive picture of big data, and can thus, generate more societal value together than each approach on its own.

Keywords Analytics · Big data · Ethnography · Thick data

2.1 Introduction

For thousands of years and across many civilizations, people have been craving for knowing the future. From asking the Oracle to consulting the crystal ball to reading the tarot cards, these activities stand as examples that show how people have always sought any help that could tell them what the future held, information that would aid them make better decisions in the present. Today, the craving for such knowledge is still alive and the means to meet it is big data and big data analytics. From traffic congestion to natural disasters, from disease outbursts to terrorist attacks, from game results to human behaviour, the general view is that there is nothing that big data

V. Charles (✉)
Buckingham Business School, University of Buckingham, Buckingham, UK
e-mail: v.charles@buckingham.ac.uk

T. Gherman
School of Business and Economics, Loughborough University, Loughborough, UK
e-mail: t.i.gherman@lboro.ac.uk

A. Emrouznejad and V. Charles (eds.), *Big Data for the Greater Good*,
Studies in Big Data 42, https://doi.org/10.1007/978-3-319-93061-9_2

analytics cannot predict. Indeed, the analysis of huge datasets has proven to have invaluable applications.

Big data analytics is one of today's most famous technological breakthroughs [1] that can enable organizations to analyse fast-growing immense volumes of varied datasets across a wide range of settings, in order to support evidence-based decision-making [2]. Over the past few years, the number of studies that have been dedicated to assessing the potential value of big data and big data analytics has been steadily increasing, which also reflects the increasing interest in the field. Organizations worldwide have come to realize that in order to remain competitive or gain a competitive advantage over their counterparts, they need to be actively mining their datasets for newer and more powerful insights.

Big data can, thus, mean big money. But there seems to be at least one problem. A 2013 survey by the big data firm Infochimps, who looked at the responses from over 300 IT department staffers, indicated that 55% of big data projects do not get completed, with many others falling short of their objectives [3]. According to another study published by Capgemini & Informatica [4], who surveyed 210 executives from five developed countries (France, Germany, Italy, the Netherlands, and the UK) to assess the business value and benefits that enterprises are realizing from big data, only 27% of big data projects were reported as profitable, whereas 45% reached their equilibrium, and 12% actually lost money. Further, in 2015, Gartner predicted that through 2017, 60% of big data projects will fail to go beyond piloting and experimentation [5]. In 2016, Gartner actually conducted an online survey among 199 Gartner Research Circle members and the results indicated that only 15% of businesses deployed their big data projects from pilot to production [6].

These statistics show that although big investments are taking place in big data projects, the generation of value does not match the expectations. The obvious question is, of course, *why*? Why are investments in big data failing, or in other words, why having the data is not sufficient to yield the expected results? In 2014, Watson [2] advanced that "the keys to success with big data analytics include a clear business need, strong committed sponsorship, alignment between the business and IT strategies, a fact-based decision-making culture, a strong data infrastructure, the right analytical tools, and people skilled in the use of analytics" (p. 1247). Although informative and without doubt useful, today nonetheless, these tips seem to be insufficient; otherwise stated, if we know what we need in order to succeed with big data analytics, then why don't we succeed in creating full value? The truth is that we are yet to profoundly understand how big data can be translated into economic and societal value [7] and the sooner we recognize this shortcoming, the sooner we can find solutions to correct it.

In this chapter, we advance that *ethnography* can support *big data analytics* in the generation of greater societal value. Although perceived to be in opposition, ethnography and big data analytics have much in common and in many ways, they have a shared purpose. In the following sections, we explore the intersection of the two approaches. Ultimately, we advance that researchers can blend big data analytics and ethnography within a research setting; hence, that big data analytics and ethnography together can inform the *greater good* to a larger extent than each approach on its own.

2.1.1 What Is Big Data?

'Big data': a concept, a trend, a mindset, an era. No unique definition, but a great potential to impact on essentially any area of our lives. The term *big data* is generally understood in terms of the four Vs advanced by Gartner [8]: volume, velocity, variety, and veracity. In time, the number of Vs has increased, reaching up to ten Vs [9]. Other authors have further expanded the scope of the definition, broadening the original framework: Charles and Gherman [10], for example, advocated for the inclusion of three Cs: context, connectedness, and complexity. One of the most elegant and comprehensive definitions of big data can be found in the none other than the Oxford English Dictionary, which defines it as: "extremely large datasets that may be analysed computationally to reveal patterns, trends, and associations, especially relating to human behaviour and interactions."

Big data comes from various structured and unstructured sources, such as archives, media, business apps, public web, social media, machine log data, sensor data, and so on. Today, almost anything we can think of produces data and almost every data point can be captured and stored. Some would say: also, analysed. This may be true, but in view of the statistics presented in the introduction above, we are reticent to so state. Undoubtedly, data is being continuously analysed for better and deeper insights, even as we speak. But current analyses are incomplete, since if were to be able to fully extract the knowledge and insights that the datasets hold, we would most probably be able to fully capitalize on their potential, and not so many big data projects would fail in the first place.

The big data era has brought many challenges with it, which deemed the traditional data processing application software unfit to deal with them. These challenges include networking, capturing data, data storage and data analysis, search, sharing, transfer, visualization, querying, updating and, more recently, information privacy [11]. But the list is not exhaustive and challenges are not static; in fact, they are dynamic, constantly mutating and diversifying. One of the aspects that we seem to generally exclude from this list of challenges is human behaviour. Maybe it is not too bold to say that one of the biggest challenges in the big data age is the extraction of insightful information not from the existing data, but from the data originating from emergent human dynamics that either haven't happened yet or that would hardly be traceable through big data.

One famous example in this regard is Nokia, a company that in the 1990s and part of 2000s was one of the largest mobile phone companies in the world, holding by 2007 a market share of 80% in the smartphone market [12]. Nevertheless, Nokia's over-dependence on quantitative data has led the company to fail in maintaining its dominance on the mobile handset market. In a post published in 2016, technology ethnographer Tricia Wang [13] describes how she conducted ethnographic research for Nokia in 2009 in China, which revealed that low-income consumers were willing to pay for more expensive smartphones; this was a great insight at the time that led her to conclude that Nokia should replace their then strategy from making smartphones for elite users to making smartphone for low-income users, as well. But Nokia considered that Wang's sample size of 100 was too small to be reliable and that moreover, her conclusion was not supported by the large datasets that Nokia possessed; they, thus, did not implement the insight. Nokia was bought by Microsoft in 2013 and Wang concluded that:

> There are many reasons for Nokia's downfall, but one of the biggest reasons that I witnessed in person was that the company over-relied on numbers. They put a higher value on quantitative data, they didn't know how to handle data that wasn't easily measurable, and that didn't show up in existing reports. What could've been their competitive intelligence ended up being their eventual downfall.

Netflix is at the other end of the game, illustrating how ethnographic insights can be used to strengthen a company's position on the market. Without doubt, Netflix is a data-driven company, just like Nokia. In fact, Netflix pays quite a lot of attention to analytics to gain insight into their customers. In 2006, Netflix launched the Netflix Prize competition, which would reward with $1 million the creation of an algorithm that would "substantially improve the accuracy of predictions about how much someone is going to enjoy a movie based on their movie preferences". But at the same time, Netflix was open to learning from more qualitative and contextual data about what users really wanted. In 2013, cultural anthropologist Grant McCracken conducted ethnographic research for Netflix and what he found was that users *really enjoyed* to watch chapter after chapter of the same series, engaging in a new form of consumption, now famously called *binge watching.* A survey conducted in the same year among 1500 TV streamers (online U.S. adults who stream TV shows at least once per week) confirmed that people did not feel guilty about binge watching, with 73% of respondents actually feeling good about it [14]. This new insight was used by Netflix to re-design its strategy and release whole seasons at once, instead of releasing one episode per week. This, in turn, changed the way users consumed media and specifically Netflix's products and how they perceived the Netflix brand, while improving Netflix's business.

2.1.2 What Is Ethnography?

Having already introduced the notion of ethnographic research in Sect. 2.1.1, let us now consider the concept of *ethnography* and discuss it further. Ethnography, from the Greek words *ethnos*, meaning 'folk, people, nation', and *grapho*, meaning 'I write' or 'writing', is the systematic study of people and cultures, aimed at understanding and making sense of social meanings, customs, rituals, and everyday practices [15, 16]. Ethnography has its origin in the work of early anthropologists, such as Bronislaw Malinowski [17] and Margaret Mead [18], who largely focused on mapping out the cultures of small and isolated tribes, before they became 'contaminated' by contact with the industrial world [19]. In time, this simple definition has been refined by many authors.

For example, Brewer [15] defined ethnography as:

> The study of people in naturally occurring settings or 'fields' by methods of data collection which capture their social meanings and ordinary activities, involving the researchers participating directly in the setting, if not also the activities, in order to collect data in a systematic manner but without meaning being imposed on them externally (p. 6).

Delamont [20], on the other hand, stated that:

> Participant observation, ethnography and fieldwork are all used interchangeably… they can all mean spending long periods watching people, coupled with talking to them about what they are doing, thinking and saying, designed to see how they understand their world (p. 218).

Ethnography is about becoming part of the settings under study. "Ethnographies are based on observational work in particular settings" (p. 37) [21], allowing researchers to "see things as those involved see things" (p. 69) [22]; "to grasp the native's point of view, his relation to life, to realize his vision of his world" (p.25) [17]. Consequently, the writing of ethnographies is viewed as an endeavour to describe 'reality' [23, 24], as this is being experienced by the people who live it.

Ethnography depends greatly on *fieldwork*. Generally, data is collected through participant or nonparticipant observation. The primary data collection technique used by ethnographers is, nonetheless, *participant observation*, wherein the researchers assume an insider role, living as much as possible with the people they investigate. Participant observers interact with the people they study, they listen to what they say and watch what they do; otherwise stated, they focus on people's doings in their natural setting, in a journey of discovery of everyday life. *Nonparticipant observation*, on the other hand, requires the researchers to adopt a more 'detached' position. The two techniques differ, thus, from one another based on the weight assigned to the activities of 'participating' and 'observing' [16, 25].

Finally, ethnography aims to be a holistic approach to the study of cultural systems [26], providing 'the big picture' and depicting the intertwining between relationships and processes; hence, it usually requires a long-term commitment and dedication. In today's fast-paced environment, however, mini-ethnographies are also possible. A mini-ethnography focuses on a specific phenomenon of interest and as such, it occurs in a much shorter period of time than that required by a full-scale ethnography [27].

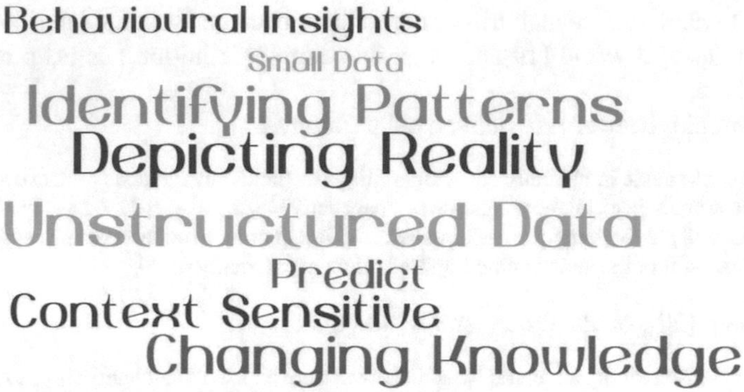

Behavioural Insights
Small Data
Identifying Patterns
Depicting Reality
Unstructured Data
Predict
Context Sensitive
Changing Knowledge

Fig. 2.1 The intersection between big data analytics and ethnography

2.2 Big Data Analytics and Ethnography: Points of Intersection

Ford [28] advanced that "data scientists and ethnographers have much in common, that their skills are complementary, and that discovering the data together rather than compartmentalizing research activities was key to their success" (p. 1). In a more recent study, Laaksonen et al. [29] postulated that "ethnographic observations can be used to contextualize the computational analysis of large datasets, while computational analysis can be applied to validate and generalize the findings made through ethnography" (p. 110). The latter further proposed a new approach to studying social interaction in an online setting, called *big-data-augmented-ethnography*, wherein they integrated ethnography with computational data collection.

To the best of our knowledge, the literature exploring the commonalities between big data analytics and ethnography is quite limited. In what follows, we attempt, thus, to contribute to the general discussion on the topic, aiming to highlight additional points of intersection. Figure 2.1 briefly depicts these points.

2.2.1 Depicting 'Reality'

Big data analytics comprise the skills and technologies for continuous iterative exploration and investigation of past events to gain insight into what has happened and what is likely to happen in the future [30]. In this sense, data scientists develop and work with models. Models, nonetheless, are simplified versions of reality. Models built aim, thus, to represent the reality and in this sense, are continuously revised, checked, and improved upon and, furthermore, tested to account for the extent to which they actually do so.

On the other hand, ethnographies are conducted in a naturalistic setting in which real people live, with the writing of ethnographies being viewed as an endeavour to describe 'reality' [23, 24]. Furthermore, just as big data analytics-informed models are continuously being revised, "ethnography entails continual observations, asking questions, making inferences, and continuing these processes until those questions have been answered with the greatest emic validity possible" (pp.17–18) [26]. In other words, both big data and ethnography are more concerned with the processes through which 'reality' is depicted rather than with judging the 'content' of such reality.

2.2.2 Changing the Definition of Knowledge

Both big data analytics and ethnography change the definition of *knowledge* and this is because both look for a more accurate representation of reality. On the one hand, big data has created a fundamental shift in how we think about research and how we define knowledge, reframing questions about the nature and the categorization of reality and having a profound change at the levels of epistemology and ethics [31]. Big data analytics offers what Lazer et al. [32] called "the capacity to collect and analyse data with an unprecedented breadth and depth and scale" (p. 722). Or as boyd and Crawford [31] wrote, "just as Du Gay and Pryke [33] note that 'accounting tools... do not simply aid the measurement of economic activity, they shape the reality they measure' (pp. 12–13), so big data stakes out new terrains of objects, methods of knowing, and definitions of social life" (p. 665).

On the other hand, ethnographies aim to provide a detailed description of the phenomena under study, and as such, they may reveal that people's reported behaviour does not necessarily match their observed behaviour. As a quote widely attributed to the famous anthropologist Margaret Mead states: "What people say, what people do, and what they say they do are entirely different things". Ethnographies can and are generally performed exactly because they can provide insights that could lead to new hypotheses or revisions of existing theory or understanding of social life.

2.2.3 Searching for Patterns

Both data scientists and ethnographers collect and work with a great deal of data and their job is, fundamentally, to identify patterns in that data. On the one hand, some say that the actual value of big data rests in helping organizations find patterns in data [34], which can further be converted into smart business insights [35]. Big data analytics or machine learning techniques help find hidden patterns and trends in big datasets, with a concern more towards the revelation of solid statistical relationships. Generally, this means finding out whether two or more variables are related or associated.

Ethnography, on the other hand, literally means to 'write about a culture' and in the course of so doing, it provides think descriptions of the phenomena under study, trying to make sense of what is going on and reveal understandings and meanings. By carefully observing and/or participating in the lives of those under study, ethnographers thus look for shared and predictable patterns in the lived human experiences: patterns of behaviour, beliefs and customs, practices, and language [36].

2.2.4 Aiming to Predict

A common application of big data analytics includes the study of data with the aim to predict and improve. The purpose of predictive analytics is to measure precisely the impact that a specific phenomenon has on people and to predict the chances of being able to duplicate that impact in future activities. In other words, identifying patterns in the data is generally used to build predictive models that will aid in the optimization of a certain outcome.

On the other hand, it is generally thought that ethnography is, at its core, descriptive. But this is somehow misunderstood. Today, there is a shift in an ethnographer's aims, whose ethnographic analyses can take the shape of predictions. Evidence of this are Wang's [13, 37] ethnographic research for Nokia and McCracken's ethnographic research for Netflix, described in Sect. 2.1.1 of this chapter. In the end, in practical terms, the reason why we study a phenomenon, irrespective of the method of data collection or data analysis used, is not just because we want to understand it better, but because we also want to predict it better. The identification of patterns enables predictions, and as we have already implied before, both big data analytics and ethnography can help in this regard.

2.2.5 Sensitive to Context

Big data analytics and ethnography are both context-sensitive; in other words, taken out of context, the insights obtained from both approaches will lose their meaning. On the one hand, big data analytics is not just about finding patterns in big data. It is not sufficient to discover that one phenomenon correlates with another; or otherwise stated, there is a big difference between identifying correlations and actually discovering that one causes the other (cause and effect relationship). Context, meaning and interpretation become a necessity, and not a luxury. This observation was also made by Maxwell [38], when he elegantly stated that:

> Analytics often happens in a black box, offering up a response without context or clear transparency around the algorithmic rules that computed a judgment, answer, or decision. Analytics software and hardware are being sold as a single- source, easy solution to make sense of today's digital complexity. The promise of these solutions is that seemingly anyone can be an analytics guru. There is real danger in this do-it-yourself approach to analyt-

ics, however. As with all scientific instruments and approaches, whether it be statistics, a microscope, or even a thermometer, without proper knowledge of the tool, expertise in the approach, and knowledge of the rules that govern the process, the results will be questionable.

On the other hand, inferences made by ethnographers are tilted towards the explanation of phenomena and relationships observed within the study group. Ethnography supports the research endeavours of understanding the multiple realities of life *in context* ([39], emphasis added); hence, by definition, ethnography provides detailed snapshots of contextualized social realities. Generalization outside the group is limited and taken out of context, meanings will also be lost.

2.2.6 'Learning' from Smaller Data

Bigger data are not always better data. Generally, big data is understood as originating from multiple sources (the 'variety' dimension) or having to be integrated from multiple sources to obtain better insights. Nevertheless, this creates additional challenges. "Every one of those sources is error-prone… […] we are just magnifying that problem [when we combine multiple datasets]" (p. 13) [40], cited by boyd and Crawford [31]. In this sense, smaller data may be more appropriate for intensive, in-depth examination to identify patterns and phenomena, an area in which ethnography holds the crown.

Furthermore, data scientists have always been searching for new or improved ways to analyse large datasets to identify patterns, but one of the challenges encountered has been that they need to know what they are looking for in order to find it, something that is particularly difficult when the purpose is to study emergent human dynamics that haven't happened yet or that will not show up that easily in the datasets. Both big data analytics and ethnography can, thus, learn from smaller datasets (or even single case analyses). On the other hand, we must acknowledge that there are also situations in which researchers generally rely on big data (such as clinical research), but sometimes they have to rely on surprisingly small datasets; this is the case of, for example, clinical drug research that analyses the data obtained after drugs are released on the market [40].

2.2.7 Presence of Behavioural Features

Le Compte and Preissle [25] once said that: "Those who study humans are themselves humans and bring to their investigations all the complexity of meaning and symbolism that complicates too precise an application of natural science procedures to examining human life" (p. 86). This has generally been presented as a difficulty that ethnographers, in particular, must fight to overcome. But it does not have to be so in the big data age. The truth is we need as many perspectives and as many insights as possible. Fear that we might go wrong in our interpretations will only stop progression. A solution to this is posed by the collaboration between data scientists and ethnographers. In this sense, ethnographers should be allowed to investigate the complexity of the data and come out with propositions and hypotheses, even when they are conflicting; and then data scientists could use big data analytics to test those propositions and hypotheses in light of statistical analyses and see if they hold across the larger datasets.

Studying human behaviour is not easy, but the truth is that both big data analytics and ethnography have a behavioural feature attached to them, in the sense that they are both interested in analysing the content and meaning of human behaviour; the 'proximity' between the two approaches is even more evident if we consider the change that the big data age has brought with it. While in the not so far past, analytics would generally be performed by means of relying upon written artefacts which recorded past human behaviour, today, big data technology enables the recording of *current* human behaviour, *as this happens* (consider live feed data, for example). And as technology will keep evolving, the necessity of 'collaboration' between big data analytics and ethnography will become more obvious.

2.2.8 Unpacking Unstructured Data

Big data includes both structured (e.g., databases, CRM systems, sales data, sensor data, and so on) and unstructured data (e.g., emails, videos, audio files, phone records, social media messages, web logs, and so on). According to a report by Cisco [41], an estimated 90% of the existing data is either semi-structured or unstructured. Furthermore, a growing proportion of unstructured data is video. And video constituted approx. 70% of all Internet traffic in 2013. One of the main challenges of big data analytics is just how to analyse all these unstructured data. Ethnography (and its newer addition, online ethnography) may have the answer.

Ethnography is a great tool to 'unpack' unstructured data. Ethnography involves an inductive and iterative research process, wherein data collection and analysis can happen simultaneously, without the need to have gathered all the data or even look at the entire data. Ethnography does not follow a linear trajectory, and this is actually an advantage that big data analytics can capitalize on. Ethnography is par excellence a very good approach to look into unstructured data and generate hypotheses that can be further tested against the entire datasets.

2.3 Final Thoughts

Today, most companies seem to be still collecting massive amounts of data with a 'just in case we need it' approach [11]. But in the wise words of William Bruce Cameron, *not everything that can be counted counts, and not everything that counts can be counted*. What should probably happen is that: before going ahead and collecting huge amounts of data, companies could use initial ethnographic observations to identify emergent patterns and phenomena of interest, advancing various, even conflicting, hypotheses. This could then inform and guide the overall strategy of massive data collection. Big data analytics could then analyse the data collected, testing the hypotheses proposed against these larger datasets. In this sense, ethnography can help shed light on the complexities of big data, with ethnographic insights serving as input for big data analytics and big data analytics can be used to generalize the findings.

Employing an ethnographic approach is generally understood in a traditional sense, which is that of having to undertake long observations from within the organization, with the researcher actually having to become an insider, a part of the organization or context that he decides to study. The good news is that today, new methods of ethnography are emerging, such as virtual ethnography (also known as online ethnography, netnography, or webnography [42]), which may turn out to be of great help in saving time and tackling the usual problem of having to gain access to the organization. The virtual world is now in its exponential growth phase and doing virtual ethnography may just be one of the best, also convenient answers to be able to explore and benefit from understanding these new online contexts. The web-based ethnographic techniques imply conducting virtual participant observation via interactions in online platforms such as social networks (such as Facebook or Twitter), blogs, discussion forums, and chat rooms. Conducting ethnographies in today's world may, thus, be easier than it seems.

In this chapter, we have aimed to discuss the points of intersection between big data analytics and ethnography, highlighting both their similarities and complementary nature. Although the list is far from being exhaustive, we hope to have contributed to the discussions that focus on how the two approaches can work together to provide a more comprehensive picture of big data. As Maxwell [38] stated, "ethnographers bring considerable skills to the table to contextualize and make greater meaning of analytics, while analytics and algorithms are presenting a new field site and complementary datasets for ethnographers" (p. 186).

One of the most important advantages of combining big data analytics and ethnography is that this 'intersection' can provide a better sense of the realities of the contexts researched, instead of treating them as abstract, reified entities. And this better sense can translate into better understandings and better predictions, which can further assist in the creation of better practical solutions, with greater societal added value. There are indeed many points of intersection between big data analytics and ethnography, having in many ways a shared purpose. They are also complementary, as data scientists working with quantitative methods could supplement their own 'hard' methodological techniques with findings and insights obtained from ethnographies. As Goodall [43] stated:

> Ethnography is not the result of a noetic experience in your backyard, nor is it a magic gift that some people have and others don't. It is the result of a lot of reading, a disciplined imagination, hard work in the field and in front of a computer, and solid research skills… (p. 10)

Today, we continue to live in a world that is being influenced by a *quantification bias,* the unconscious belief of valuing the measurable over the immeasurable [37]. We believe it is important we understood that big data analytics has never been one size fits all. We mentioned in this chapter that many big data projects fail, despite the enormous investments that they absorb. This is in part because many people still fail to comprehend that a deep understanding of the context in which a pattern emerge is not an option, but a must. Just because two variables are correlated does not necessarily mean that there is a cause and effect relationship taking place between them. boyd and Crawford [31] meant exactly that when they stated that:

> Too often, Big Data enables the practice of apophenia: seeing patterns where none actually exist, simply because enormous quantities of data can offer connections that radiate in all directions. In one notable example, Leinweber [44] demonstrated that data mining techniques could show a strong but spurious correlation between the changes in the S&P 500 stock index and butter production in Bangladesh.

Ethnography can provide that so very necessary deep understanding. In this sense, big data analytics and ethnography can work together, complementing each other and helping in the successful handcrafting and implementation of bigger projects for a bigger, greater good.

References

1. R.G. Fichman, B.L. Dos Santos, Z. Zheng, Digital innovation as a fundamental and powerful concept in the information systems curriculum. MIS Q. **38**(2), 329–353 (2014)
2. H.J. Watson, Tutorial: big data analytics: concepts, technologies, and applications. Commun. Assoc. Inf. Syst. **34**(1), 1247–1268 (2014)
3. Infochimps, CIOs & Big Data: What Your IT Team Wants You to Know. [online] (2013). Whitepaper. http://www.infochimps.com/resources/report-cios-big-data-what-your-it-team-wants-you-to-know-6/. Accessed 11 October 2017
4. Capgemini & Informatica, The big data payoff: turning big data into business value. A joint report by Informatica and Capgemini on the keys to operationalizing Big Data projects. [online] (2016). https://www.capgemini.com/de-de/wp-content/uploads/sites/5/2017/07/the_big_data_payoff_turning_big_data_into_business_value.pdf. Accessed 15 October 2017
5. Gartner, Gartner Says Business Intelligence and Analytics Leaders Must Focus on Mindsets and Culture to Kick Start Advanced Analytics. (2015). Gartner Business Intelligence & Analytics Summit 2015, 14–15 October 2015, Munich, Germany. [online] September (2015). https://www.gartner.com/newsroom/id/3130017. Accessed 20 October 2017
6. Gartner, Gartner Survey Reveals Investment in Big Data Is Up but Fewer Organizations Plan to Invest. Gartner Business Intelligence & Analytics Summit 2016, 10–11 October 2016, Munich, Germany. [online] October (2016). https://www.gartner.com/newsroom/id/3466117. Accessed 11 October 2017
7. W.A. Günther, M.H. Rezazade Mehrizi, M. Huysman, F. Feldberg, Debating big data: a literature review on realizing value from big data. J. Strateg. Inf. Syst. **26**, 191–209 (2017)
8. D. Laney, 3D Data Management: controlling data volume, velocity and variety. Applications delivery strategies, META Group (now Gartner) [online] (2001). http://blogs.gartner.com/ doug-laney/files/2012/01/ad949-3D-Data-Management-Controlling-Data-Volume-Velocity-and-Variety.pdf. Accessed 1 Aug 2017

9. L. Markus, New games, new rules, new scoreboards: the potential consequences of big data. J. Inf. Technol. **30**(1), 58–59 (2015)
10. V. Charles, T. Gherman, Achieving competitive advantage through big data. Strategic implications. Middle-East J. Sci. Res. **16**(8), 1069–1074 (2013)
11. V. Charles, M. Tavana, T. Gherman, The right to be forgotten—is privacy sold out in the big data age? Int. J. Soc. Syst. Sci. **7**(4), 283–298 (2015)
12. H. Bouwman et al., How Nokia failed to nail the smartphone market. Conference Paper. 25th European Regional Conference of the International Telecommunications Society (ITS), 22–25 June 2014, Brussels, Belgium (2014)
13. T. Wang, Why big data needs thick data. [online] January (2016a). https://medium.com/ethnography-matters/why-big-data-needs-thick-data-b4b3e75e3d7. Accessed 13 December 2017
14. Netflix Media Center, Netflix Declares Binge Watching is the New Normal. [online] December (2013). https://media.netflix.com/en/press-releases/netflix-declares-binge-watching-is-the-new-normal-migration-1. Accessed 5 January 2018
15. J.D. Brewer, *Ethnography* (Open University Press, Philadelphia, PA, 2000)
16. R. Madden, *Being Ethnographic: A Guide to the Theory and Practice of Ethnography* (Sage, London, UK, 2010)
17. B. Malinowski, *Argonauts of the Western Pacific* (Routledge and Kegan Paul, London, 1922)
18. M. Mead, *Coming of Age in Samoa: A Study of Adolescence and Sex in Primitive Societies* (Penguin, Harmondsworth, 1943)
19. M. Descombe, *The Good Research Guide for Small-scale Social Research Projects*, 3rd edn. (Open University Press, McGraw-Hill, Berkshire, UK, 2007)
20. S. Delamont, Ethnography and participant observation, in *Qualitative Research Practice*, ed. by C. Seale, G. Gobo, J. Gubrium, D. Silverman (Sage, London, 2004), pp. 217–229
21. D. Silverman, *Doing Qualitative Research. A Practical Handbook* (Sage, Thousand Oaks, CA, 2000)
22. M. Descombe, *The Good Research Guide for Small-scale Social Research Projects* (Open University Press, Buckingham, 1998)
23. M. Hammersley, P. Atkinson, *Ethnography: Principles in Practice* (Tavistock, New York, 1983)
24. D. Silverman, *Interpreting Qualitative Data*, 2nd edn. (Sage, London, 2001)
25. M.D. LeCompte, J. Preissle, *Ethnography and Qualitative Design in Educational Research*, 2nd edn. (Academic Press Inc., San Diego, CA, 1993)
26. T.L. Whitehead, What is Ethnography? Methodological, Ontological, and Epistemological Attributes. Ethnographically Informed Community and Cultural Assessment Research Systems (EICCARS) Working Paper Series. [online] (2004). Cultural Ecology of Health and Change (CEHC). http://www.cusag.umd.edu/documents/workingpapers/epiontattrib.pdf. Accessed 8 September 2017
27. K.L. White, Meztizaje and remembering in Afro-Mexican communities of the Costa Chica: Implications for archival education in Mexico. Arch. Sci. **9**, 43–55 (2009)
28. H. Ford, Big data and small: Collaborations between ethnographers and data scientists. (2014). Big Data & Society, July–December, 1–3
29. S.M. Laaksonen, M. Nelimarkka, M. Tuokko, M. Marttila, A. Kekkonen, M. Villi, Working the fields of big data: using big-data-augmented online ethnography to study candidate–candidate interaction at election time. J. Inf. Technol. Politics **14**(2), 110–131 (2017)
30. J. Mustafi, Natural language processing and machine learning for big data, in ed. by B.S.P. Mishra, S. Dehuri, E. Kim, G-N., *Wang Techniques and Environments for Big Data Analysis. Parallel, Cloud, and Grid Computing*. Studies in Big Data 17 (Springer, Switzerland, 2016), pp. 53–74
31. D. boyd, K. Crawford, Critical questions for big data: provocations for a cultural, technological, and scholarly phenomenon. Inf. Commun. Soc. **15**(5), 662–679 (2012)
32. D. Lazer et al., Computational social science. Science **323**, 721–723 (2009)

33. P. Du Gay, M. Pryke, *Cultural Economy: Cultural Analysis and Commercial Life* (Sage, London, UK, 2002)
34. B. Hayes, Big data relationships. Blog. Big Data & Analytics Hub. [online] July (2014). http://www.ibmbigdatahub.com/blog/big-data-relationships. Accessed 10 September 2017
35. E. Brynjolfsson, A. McAfee, Big data: the management revolution. Harvard Bus. Rev. **90**(10), 60–68 (2012)
36. M. Angrosino, *Doing Ethnographic and Observational Research* (Sage, Thousand Oaks, CA, 2007)
37. T. Wang, The human insights missing from big data. TEDxCambridge. [online] September (2016b). https://www.ted.com/talks/tricia_wang_the_human_insights_missing_from_big_data. Accessed 1 December 2017
38. C.R. Maxwell, Accelerated pattern recognition, ethnography, and the era of big data, ed. by B. Jordan, *Advancing Ethnography in Corporate Environments. Challenges and Emerging Opportunities* (Walnut Crest, CA: Left Coast Press, 2013), pp. 175–192
39. G.B. Rossman, S.F. Rallis, *Learning in the Field: An Introduction to Qualitative Research* (Sage, London, 2003)
40. D. Bollier, The promise and peril of big data. [online] (2010). https://assets.aspeninstitute.org/content/uploads/files/content/docs/pubs/The_Promise_and_Peril_of_Big_Data.pdf. Accessed 15 September 2017
41. Cisco, Big data: not just big, but different. [online] (2014). https://www.cisco.com/c/dam/en_us/about/ciscoitatwork/enterprise-networks/docs/i-bd-04212014-not-just-big-different.pdf. Accessed 30 December 2017
42. A. Purli, The web of insights: the art and practice of webnography. Int. J. Market Res. **49**, 387–408 (2007)
43. H.L. Goodall, *Writing the New Ethnography* (AltaMira, Lanham, MA, 2000)
44. D. Leinweber, Stupid data miner tricks: overfitting the S&P 500. J. Invest. **16**(1), 15–22 (2007)

Prof. V. Charles is the Director of Research and Professor of Management Science at Buckingham Business School, The University of Buckingham, UK. He is an Honorary Professor at MSM, The Netherlands; an Affiliated Distinguished Research Professor at CENTRUM Católica GBS, Peru; a Fellow of the Pan-Pacific Business Association, USA; a Council Member of the Global Urban Competitiveness Project, USA; an HETL Liaison to Peru, HETL USA; and a WiDS Ambassador, WiDS Buckingham (in collaboration with Stanford University, USA). He is a PDRF from NUS, Singapore and a certified Six-Sigma MBB. He holds executive certificates from the HBS, MIT, and IE Business School. He has 20 years of teaching, research, and consultancy experience in various countries, in the fields of applied quantitative analytics (big data). He has over 120 publications. He holds international honours/awards, among which the 2016 Cum Laude Award from the Peruvian Marketing Society; the most innovative study to Peru's Think-Tank of the Year 2013; the highly commended award of Emerald/CLADEA and the intellectual contribution award of CENTRUM for three and seven consecutive years, respectively. He has developed the 4^3 Architecture for Service Innovation. His area of focus includes productivity, quality, efficiency, effectiveness, competitiveness, innovation, and design thinking.

Tatiana Gherman graduated as an Economist, with a B.A. degree in International Business and Economics from the Academy of Economic Studies of Bucharest, Romania. She then underwent her MBA with a full scholarship at CENTRUM Católica Graduate Business School in Lima, Peru, where she concluded the programme CUM LAUDE, as an MBA topper, and was awarded the first prize. Her academic achievement has been recognised and awarded by the International Honor Society Beta Gamma Sigma. She is currently a doctoral research student at the School of Business and Economics, Loughborough University, UK, where she is investigating the area of behavioural operational research. Her research interests include the application of ethnography, ethnomethodology, and conversation analysis to the study of the practice of management science/operational research; multi-attribute decision-making techniques (The Technique for Order of Preference by Similarity to Ideal Solution and Analytical Hierarchy Process); and other quantitative analytics at different levels.

Chapter 3
Big Data: A Global Overview

Celia Satiko Ishikiriyama and Carlos Francisco Simoes Gomes

Abstract More and more, society is learning how to live in a digital world that is becoming engulfed in data. Companies and organizations need to manage and deal with their data growth in a way that compliments the data getting bigger, faster and exponentially more voluminous. They must also learn to deal with data in new and different unstructured forms. This phenomenon is called Big Data. This chapter aims to present other definitions for Big Data, as well as technologies, analysis techniques, issues, challenges and trends related to Big Data. It also looks at the role and profile of the Data Scientist, in reference to functionality, academic background and required skills. The result is a global overview of what Big Data is, and how this new form is leading the world towards a new way of social construction, consumption and processes.

3.1 Big Data

3.1.1 The Origins of Big Data and How It Is Defined

From an evolutionary perspective, Big Data is not new [58]. The advance towards Big Data is a continuation of ancient humanity's search for measuring, recording and analyzing the world [42]. A number of companies have been using their data and analytics for decades [17].

The most common and widespread definition for Big Data refers to the 3 Vs: volume, velocity and variety. Originally, the 3Vs were pointed out by Doug Laney in 2001, in a Meta Group report. In this report, Laney [35] identifies the 3Vs as future challenges in data management and is nowadays widely used to define Big Data [22].

C. S. Ishikiriyama (✉) · C. F. S. Gomes
Universidade Federal Fluminense, Niteroi, Brazil
e-mail: csatiko@gmail.com

C. F. S. Gomes
e-mail: cfsg1@bol.com.br

A. Emrouznejad and V. Charles (eds.), *Big Data for the Greater Good*,
Studies in Big Data 42, https://doi.org/10.1007/978-3-319-93061-9_3

Although the 3Vs are the most solid definition for Big Data, they are definitely not the only one. Many authors have attempted to define and explain Big Data under a number of perspectives, going through more detailed definitions—including technologies and data analysis techniques, the use and goals of Big Data and also the transformations it is imposing within industries, services and lives.

The expression Big Data and Analytics (BD&A) has become synonymous with Business Intelligence (BI) among some suppliers and for others, BD&A was an incorporation of the traditional BI but with the addition of new elements such as predictive analyses, data mining, operation tools/approaches and also research and science [23]. Reyes defines Big Data as the process of assembling, analyzing and reporting data and information [47].

Big Data and BD&A are often described as data sets and analytical techniques in voluminous and complex applications that require storage, management, analysis and unique and specific visualization technologies [9]. They also include autonomous data sources with distributed and decentralized controls [60].

Big Data has also been used to describe a large availability of digital data and financial transactions, social networks and data generated by smartphones [41]. It includes non-structured data with the need for real-time analysis [10]. Although one of Big Data's main characteristics is the data volume, the size of data must be relative, depending on the available resources as well as the type of data that is being processed [33].

Mayer-Schonberger and Cukier believe that Big Data refers to the extraction of new ideas and new ways to generate value in order to change markets, organizations, the relationship between citizens and government, and so on. It also refers to the ability of an organization to obtain information in new ways, aiming to generate useful ideas and significant services [42].

Although the 3 Vs' characteristics are intensely present in Big Data definitions throughout literature, its concept gained a wider meaning. A number of characteristics are related to Big Data, in terms of data source, technologies and analysis techniques, goals and generation of value.

In summary, Big Data are enormous datasets composed by both structured and non-structured data, often with the need for real-time analysis and use of complex technologies and applications to store, process, analyze and visualize information from multiple sources. It plays a paramount role in the decision making process in the value chain within organizations.

3.1.2 What Are the Goals of Big Data?

Big Data promises to fulfill the research principles of information systems, which is to provide the right information for the right use, in the precise volume and quality at the right time [48]. The goal of BI&A is to generate new knowledge (insights) that can be significant, often in real-time, complementing traditional statistics research and data source's files that remain permanently static [54].

Big Data can make organizations more efficient through improvements in their operations, facilitating innovation and adaptability and optimizing resource allocation [34]. The ability of crossing and relating private data about products and consumer preferences with information from tweets, blogs, product analysis and social network data, open various possibilities for companies to analyze and understand the preferences and needs of the customers, predict demand and optimize resources [5].

The key to extracting value from Big Data is the use of Analytics, since the collection and storage themselves add little value. Data needs to be analyzed and its results used by decision makers and organizational process [58].

The emergence of Big Data is creating a new generation of data for decision support and management and is launching a new area of practice and study called Data Science. It encompasses techniques, tools, technologies and processes to extract reason out of Big Data [58]. Data Science refers to qualitative and quantitative applications to solve relevant problems and predict outputs [57].

There are a number of areas that can be impacted by the use of Big Data. Some of them include business, sciences, engineering, education, health and society [51]. Within education, some examples of Big Data application are: tertiary education management and institutional applications (including recruitment and admission processes), financial planning, donator tracking and monitoring student performance [45].

A summary of the main concepts associated with Big Data's goals, as well as the means to achieve those goals and main areas of application can be seen in Table 3.1.

3.1.3 What Is Big Data Transforming in the Data Analysis Sector?

Three big changes that represent a paradigm shift were pointed out by Mayer-Schonberger and Cukier [42]. The first of them is that the need for samples was due to a time where information was something limited. The second is that the obsession for correct data and the concern for the quality of the data were due to the short availability of data. The last is the abandonment of the search for causality and contentment and to shift focus to the discovery of the fact itself [42].

For the first big change, the argument is based on the Big Data definition itself, meaning in relative terms and not absolute. It was unviable and expensive to study a whole universe and is reinforced by the fact that nowadays some companies collect as much data as possible [42].

The second big change refers to the obsession for correct data, which adds to the first change: data availability. Before, there was limited data, so it was very important to ensure the total quality of the data. The increase of data availability opened the doors to inaccuracy and Big Data transforms the numbers into something

Table 3.1 Goals, means and applications areas of Big Data

Goals	Decision support
	Demand forecasting
	Efficiency
	Innovation
	New knowledge (insights)
	Resource optimization
	Value generation
Means	Processes
	Statistics
	Techniques
	Technologies
Application Areas	Business
	Education
	Electronic commerce
	Engineering
	Financial planning
	Government
	Health
	Recruitment
	Sciences
	Social networks

more probabilistic than precise [42]. That is, the larger the scale, more accuracy is lost [26].

Finally, the third big change in the Big Data era is that the predictions based on correlations are in Big Data's essence. That means that Big Data launches non causal analyses in a way to transform the way the world is understood. The mentality has changed on how data could be used [42].

The three changes described above turn some traditional perspectives of data analysis upside down, concerning not only the need of sampling or data quality but also integrity. It goes further, when a new way to look at data and what information to extract from it is brought to the table.

3.2 Information Technologies

The role of IT (Information Technology) in the Big Data area is fundamental and advances that occurred in this context made the arrival of this new data-driven era possible. New Big Data technologies are enabling large scale analysis of varied data, in unprecedented velocity and scale [8].

Typical sources of Big Data can be classified from the perspective of how they were generated, as follows [11]:

- User generated content (UGCs) e.g. blogs, tweets and forum content;
- Transactional data generated by large scale systems e.g. web logs, business transactions and sensors;
- Scientific data from data intensive experiments e.g. celestial data or genomes;
- Internet data that is collected and processed to support applications;
- Chart data composed by an enormous number of nodes of information and the relationship between them.

In the Information Technology industry as a whole, the speed that Big Data appeared generated new issues and challenges in reference to data and analytical management. Big Data technology aims to minimize the need for hardware and reduce processing costs [30]. Conventional data technologies, such as data bases and data warehouses, are becoming inadequate for the amount of data to analyze [11].

Big Data is creating a paradigm change in the data architecture, in a way that organizations are changing the way that data is brought from the traditional use of servers to pushing computing to distributed data [59]. The necessity for Big Data analysis boosted the development of new technologies. In order to permit processing so much data, new technologies emerged, like MapReduce from Google and its open source equivalent Hadoop, launched by Yahoo [42].

The MapReduce technology allows the development of approaches that enable the handling of a large volume of data using a big number of processors, resulting in directing for some of the problems caused by volume and velocity [43].

Apache Hadoop is one of the software's platforms that support data application in a distributed and intensive way and implement Map/Reduce [12]. Hadoop is an open source project hosted by the Apache Software Foundation and consists of small subprojects and belongs to the infra-structure category of distributed computing [29].

The role of IT in the information flow's availability to create competitive advantages was identified and pointed out as six components [24]:

- Add volume and growth, through improvement or development of products and services, channels or clients;
- Distinguish or increase the will to pay;
- Reduce costs;
- Optimize risks and operations;
- Improve industry structure, innovate with products or services and generate and make knowledge and other resources and competencies available;
- Transform models and businesses processes to continuous relevance in the scenario changes.

Cloud computing is a key component for Big Data, not only because it provides infra-structure and tools, but also because it is a business model that BD&A can trace, as it is offered as a service (Big Data as a Service—BdaaS). However, it brings a lot of challenges [5].

An intensive research project using academic papers about Big Data showed the following technologies as the most cited by the authors, by order of relevance: Hadoop/MapReduce, NoSQL, In-Memory, Stream Mining and Complex Event Processing [27].

3.3 Analysis Techniques

Normally, Big Data refers to large amounts of complex data and the data is often generated in a continuous way, implying that the data analysis occurs in real-time. Classical analysis techniques are not enough and end up being replaced by learning machine techniques [2].

Big Data's analysis techniques encompasses various disciplines, which include statistics, data mining, machine learning, neural networks, social network analysis, sign processing, pattern recognition, optimization methods and visualization approaches [12]. In addition to new processing and data storage technologies, programming languages like Python and R gained importance.

Modeling decision methods also include discrete simulation, finite elements analysis, stochastic techniques, and genetic algorithms among others. Real-time modeling is not only concerned about time and algorithm output, it is the type of work that requires additional research [51].

The opportunities of emerging analytical research can be classified in five critical technical areas: BD&A, text analytics, web analytics, social network analytics and mobile analytics [9]. Some sets of techniques receive special names, based on the way the data was obtained and the type of data to be analyzed [16], as follows:

- Text Mining: techniques to extract information from textual data, which involves statistical analysis, machine learning and linguistics;
- Audio Analytics: non-structured audio data analyses, also known as speech analytics;
- Video Analytics: encompasses a variety of techniques to monitor, analyze and extract significant information out of video transmissions;
- Social Network Analytics: analysis of both structured and non-structured data from social networks;
- Predictive Analytics: embraces a number of techniques to predict future results based on historical and current data and can be applied to most disciplines.

Besides the data analysis techniques, visualization techniques are also fundamental in this discipline. Big Data is a study of transforming data, information and knowledge in an interactive visual representation [39].

Under the influence of Big Data's technologies and techniques (large scale data mining, time series analysis and pattern mining), data like occurrences and logs can be captured in a low granularity with a long history and analyzed in multiple projections [7].

The analysis and exploration of datasets made analysis directed towards data (data-driven) possible and presents the potential to argue or even replace ad hoc analysis, for other types of analysis: consumer behavior tracking, simulations and scientific experiments and validation of hypothesis [3].

3.4 The Practical Use of the Big Data

As seen in Sect. 3.1.2, Big Data has practical applications in multiple areas: commerce, education, business, financial institutions and engineering, among many others. The Data Scientist, using programming skills, technology and data analysis techniques presented in the last section, will support the decision making process, provide insights and generate value for businesses and organizations.

The use of data supports the decision makers when responding to challenges [32]. Moreover, understanding and using Big Data improves the traditional way of the decision making process [19]. Big data can assist not only the expansion of products and services, but also enables the creation of new ones [14]. The use of Big Data is not limited to the private sector; it shows great potential in public administration. In this section, some examples of use in both spheres for the greater good are presented.

Some businesses, due to the volume of generated data, might find Big Data to be of more use to improve processes, monitor tasks and gain competitive advantages. Call centers, for instance, have the opportunity of analyzing the audio of calls which will help to both control business processes—by monitoring the agent behavior and liability—and to improve the business, having the knowledge to make the customer experience better and identifying issues referring to products and services [22].

Although predictive analytics can be used in nearly all disciplines [22], retailers and online companies are big beneficiaries of this technique. Due to the large amount of transaction operations happening every day, a number of different opportunities for insights are possible, including: understanding the behavior of the customers and consumption patterns, knowing what their customers like to buy and when, predicting sales for better sales planning and replenishment and analyzing promotions are just a few examples of what Big Data can add.

Not only has the private sector experiences the benefits of Big Data. Opportunities in public administration are akin to private organizations [40]. Governments use Big Data to stimulate the public good in the public sphere [31], by digitizing administrative data, collecting and storing more data from multiple devices [15].

Big Data can be used in the different functions of the public administration: to detect irregularities, for general observation of regulated areas, to understand the social impact through social feedback on actions taken and also to improve public services [40]. Other examples of good use of Big Data in public administration are to identify and address basic needs in a faster way, to reduce the unemployment rate, to avoid delays in pension payments, to control traffic using live streaming data and also to monitor the potential need of emergency facilities [4].

Companies and public organizations have been using their data not only to understand the past, but also to understand what is happening now and what will happen in the future. Some decisions are now automated by algorithms and findings that would have taken months or even years before are being discovered at a glance now. This great power enables a much faster reaction to situations and is leading us into a new evolution of data pacing, velocity and understanding.

3.5 The Role and Profile of the Data Scientist

With the onset of the Big Data phenomenon, there emerges the need for skilled professionals to perform the various roles that the new approach requires: the Data Scientist. In order to understand the profile of this increasingly important professional, it is paramount to understand the role that they perform in Big Data.

Working with Big Data encompasses a set of different abilities from the ones organizations are used to. Because of that, it is necessary to pay attention to a key of success for this kind of project: people. Data Scientists are necessary for Big Data to make sense [28].

The managers of organizations need to learn what to do with new data sources. Some of them are willing to hire Data Scientists with high income to work in a magical way. First, they need to understand the Data Scientist's purpose and why is it necessary to have someone playing this role [44].

The role of the Data Scientist is to discover patterns and relationships that have never been thought of or seen before. These findings must be transformed into information that can be used to take actions and generate value to the organization [58]. Data Scientists are people that understand how to fish answers for important business questions given exorbitant non-structured information [13].

Data Scientists are highly trained and curious professionals with a taste for solving hard problems and a high level of education (often Ph.D.) in analytical areas such as statistics, operational research, computer science and mathematics [58]. Statistics and computing are together the main technologies of Data Science [25].

Data Science encompasses much more than algorithms and data mining. Successful Data Scientists must be able to visualize business problems in the data perspective. There is a thinking structure of data analysis and basic principles that must be understood [46]. The most basic universal ability of the Data Scientist is to write programming codes, although the most dominant characteristic of the Data Scientist is an intense curiosity. Data Scientists are a hybrid of hacker, analyst, communicator and trust counselor.

Overall, it is necessary to think of Big Data not only in analytical terms, but also in terms of developing high level skills that enable the use of the new generation of IT tools and data collect architectures. Data must be collected from several sources, stored, organized, extracted, and analyzed in order to generate valuable findings. These discoveries must be shared with the main actors of the organization who are looking to generate competitive advantage [56].

Analytics is a complex process that demands people with a very specific educational specialization and this is why tools are fundamental to help people to execute tasks [5]. Tools and computer programming skills, including Python and R, knowledge in MapReduce and Hadoop to process large datasets; machine learning and a number of other visualization tools: Google Fusion Tables, Infogram, Many Eyes, Statwing, Tableau Public and DataHero [1].

Big Data intensifies the need for sophisticated statistics and analytical skills [59]. With all their technical and analytical skills, Data Scientists are also required to have solid domain knowledge. In both contexts, a consistent time investment is required [57]. In summary, Data Scientists need to gather a rich set of abilities, as follows [58]:

- Understand the different types of data and how they can be stored;
- Computer programming;
- Data access;
- Data analysis;
- Communicate the findings through business reports.

For Data Science to work a team of professionals with different abilities is necessary and Data Science's projects shall not be restricted to data experiments. Besides that, it is necessary to connect the Data Scientist to the world of the business expert [53]. It is very common for Data Scientists to work close to people from the organization that have domain knowledge of the business [58].

Because of this, it is useful to consider analytical users on one side and data scientists and analysts on the other side. Each group needs to have different capabilities, including a mixture of business, data and analytical expertise [58]. Analytical talents can be divided in three different types [52]:

- Specialists—that processes analytical models and algorithms, generate results and present the information in a way that organizational leaders can interpret and act;
- Experts—which are in charge of developing sophisticated models and apply them to solve business questions;
- Scientists—who lead the expert team/specialists and are in charge of constructing a story, creating innovative approaches to analyze data and producing solutions. Such solutions will be transformed into actions to support organizational strategies.

In Table 3.2, there is a summary of the Data Scientist profile, work purpose, educational level, knowledge areas and technical and non-technical skills.

Having a very wide and yet specific profile, a mixture of technique and business knowledge, the Data Scientist is a rare professional. The difficulty in finding people with the technical abilities to use Big Data tools has not gone unnoticed by the media [50]. Because of all those requirements, Data Scientists are not only limited but also expensive [58].

Table 3.2 Data Scientist's profile

Data Scientist's profile	
Purpose	Pattern discoveries never seen before, solve business problems, generate competitive advantage, support decision for actions, operational improvements and direct strategic plans.
Educational level	High educational level, often Ph.D.
Knowledge area	Computer Science
	Statistics
	Operational Research
	Mathematics
Technical skills	Database queries
	Machine Learning
	Optimization algorithms
	Programming languages
	Quantitative statistical analysis
	Simulation
	Visualization
Non-technical skills	Communication
	Curiosity
	Domain knowledge
	Taste for difficult problems

3.6 Issues, Challenges and Trends

3.6.1 Issues and Challenges

The different forms of data, ubiquity and dynamic nature of resources are a big challenge. In addition, the long reach of data, findings, access, processing, integration and physical world interpretation through data are also challenging tasks [6]. Big Data characteristics are intimately connected to privacy, security and consumer well-being and have attracted the attention of schools, businesses and the political sphere [34].

Several challenges and issues involving Big Data have arisen, not only in the context of technology or management issues but also legal matters [29]. The following issues will be discussed in this section: user privacy and security, risk of discrimination, data access and information sharing, data storage and processing capacity, analytical issues, skilled professionals, processes changing, marketing, the Internet of Things (IoT) and finally, technical challenges, which seems to be one of the issues with the most concern in the literature.

The first issue refers to privacy and security. Personal information combined with other data sources can infer other facts about one person that may be a secret or not wanted to be shared by the user. User's information is collected and used to add more value to a business or organization, many times without being aware that their personal data is being analyzed [29].

The privacy issue is particularly relevant, since there is data sharing between industries and for investigative purposes. That goes against the principle of privacy, which refers to avoiding data reutilization. The advances in BD&A provided tools to extract and correlate data, enabling privacy violations easier [8]. Preventing data access is also important for security matters against cybernetic attacks and enabling criminals to know more about their target [21].

Besides privacy matters, Big Data applications may generate concerning ethical preoccupations like social injustice or even discriminatory procedures, such as removing job possibilities to certain people, health access or even changing the social and economic level in a particular group [41].

On one hand, a person can obtain advantages from predictive analysis yet someone else may be disadvantaged against. Big Data used for law applications increase the chances that one person suffers consequences, without having the right to object or even further, without having the knowledge that they are being discriminated against [29].

Issues about data access and information sharing refer to the fact that data is used for precise decision making at the right time. For that, data needs to be available at the perfect time and in a complete manner. These demands make the process of management and governance very complex, with the additional need to make this data available for government agencies in a specific pattern [29].

Another issue about Big Data refers to storage and processing. The storage capacity is not enough for the amount of data being produced: social media websites are the major contributors, as well as sensors. Due to the big demand, outsourcing data to the cloud can be an option but loading all this data does not resolve the problem since Big Data needs to relate data and extract information. Besides the time of data uploading, data changes very rapidly, making it even harder to upload data in real-time [29].

Analytical challenges are also posed in the Big Data context. A few questions need an answer: What if the volume is so big that is not known how to deal with it? Does all data need to be stored? Does all data need to be analyzed? How to figure out what are the most relevant points? How can data bring more advantages [29]?

As seen in the last session, the Data Scientist profile is not easy. Required skills are still at an early stage. With emerging technologies, Data Science will have to be appealing to organizations and youth with a number of abilities. These skills must not be limited to technical abilities but also must extend to research, analytics, data interpreting and creativeness. These skills require training programs and the attention of universities to include Big Data in their courses [29].

The shortage of Data Scientists is becoming a serious limitation in some sectors [13]. Universities and educational institutions must offer courses capable of providing all this knowledge for a new generation of Data Scientists. University students should have enough technical abilities to conduct predictive analysis, statistical techniques knowledge and handling tools available [24].

The Master of Science curriculum should have more emphasis in what concerns Business Intelligence and Business Analytics techniques and in application development, using high level technology tools to solve important business problems [59]. Another challenge here is how fast the universities can make their course updated with so many new technologies appearing every day.

Challenges referring to issues associated with change and the implementation of new processes and even business models, especially in reference to all the data made available through the internet and also in the marketing context. The digital revolution in society and marketing has created huge challenges for companies, which encompasses discussions on the effects of sales and business models, consequences to new digital channels and media with the prevailing data growth [37].

The four main challenges for marketing are [37]:

1. The use of customer's insights and data to compete in an efficient way
2. The Power of social media for brands and customer relationships
3. New digital metrics and effective evaluation of digital marketing activities
4. The growing gap of talents with analytical capabilities in the companies.

In the context of the IoT or Internet of Everything (IoE), the following challenges are posed [18]:

- Learn the maturity capacity in terms of technologies and IT;
- Understand the different types of functionalities of IoT, that can be incorporated and how it will impact the value of the client;
- Comprehend the role of machine learning and predictive analytical models;
- Rethink business models and the value chain, based on the velocity of market change and relative responsiveness of the competition.

Finally, the technical challenges refer to error tolerance, scalability, data quality, and the need for new platforms and tools. With the arrival of the new technology, an error must be acceptable or the task must be restarted. Some of the methods of Big Data computing tend to increase the error tolerance and reduce the efforts to restart a certain task [29].

The scalability issue already took computing to the cloud, which aggregates loads of work with varied performance in large groups, requiring a high level of resource sharing. These factors combine to bring a new concern of how to program, even in complex tasks of machine learning [29].

Collecting and storing a massive amount of data come at a price. As more data drives the decision making or predictive analysis in the business, this will lead to better results. That generates some issues regarding relevant, quantity, data precision and obtained conclusions [29]. The issue of data origin is another challenge, as Big Data allows data collection from different sources to make data validation hard [8].

New tools and analytical platforms are required to solve complex optimization problems, to support the visualization of large sets of data and how they relate to each other and to explore and automate multifaceted decisions in real-time [59].

Some new modern laws of data protection make it possible for a person to find out which information is being stored, but everyone should know when an organization is collecting data and with which purposes, if it is going to be available to third parties and the consequences of not supplying the information [20].

Big Data has brought challenges in so many senses and also in so many unexpected ways. In this commotion, it is also an important issue for Big Data if the companies or organizations measuring and perceiving the return on investment (ROI) on its implementation [49].

3.6.2 Trends

BD&A applications for institutional purposes are still in their early stage and will take years to become mature, although its presence is already perceived and shall be considered [45]. The future of Big Data might be the same quantitatively, but bigger and better or disruptive forces may occur which will change the whole computing outlook [38].

The term Big Data as we know today may be inappropriate in the future, as it changes with technologies available and computing capabilities, the trend is that the scale of Big Data we know today will be small in ten years [43].

In the era of data, these become the most valuable good and important to the organizations, and may become the biggest exchange commodity in the future [55]. Data is being called "the new oil", which means that they are being refined and becoming high valued products, through analytical capabilities. Organizations need to invest in constructing this infrastructure now so they are prepared when supply and value chains are transformed [18].

An online paper was published in February, 2015 on the Forbes website and signed by Douglas Lane, which presented the three big trends for business intelligence for the next year, boosted by the use of massive data volume. The first of them says that up to 2020, information will be used to reinvent, digitalize or eliminate 80% of business processes and products from the previous decade [36].

With the growth of the IoT, connected devices, sensors and intelligent machines, the ability of things to generate new types of information in real-time also grows and will actively participate in the industry's value chain. Laney states that things will become self-agents, for people and for business [36].

The second trend is that by 2017, more than 30% of Big Data companies access will occur through data services brokers as intermediates, offering a base for businesses to make better decisions. Laney projects the arrival of a new category of business centered in the cloud which will deliver data to be used in the business connect, with or without human intervention [36].

Finally, the last trend is that "…by 2017, more than 20% of customer-facing analytic deployments will provide product tracking information leveraging the IoT". The rapid dissemination of the IoT will create a new style of customer analysis and product tracking, utilizing the ever cheapening electronic sensors, which will be incorporated in all sorts of products [36].

References

1. A. Affelt, Acting on big data a data scientist role for info pros. Online Search **38**(5), 10–14 (2014)
2. D. Agrawal, Analytics based decision making. J. Indian Bus. Search **6**, 332–340 (2014)
3. A. Alexandrov, R. Bergmann, S. Ewen et al., The Stratosphere platform for big data analytics. VLDB J. **23**, 939–964 (2014)
4. J. Archenaa, E.A. Mary Anita, A survey of big data analytics in healthcare and government. Procedia Comput. Sci. **50**, 408–413 (2015)
5. M.D. Assunção, R.N. Calheiros, S. Bianchia et al., Big Data computing and clouds: trends and future directions. Parallel Distrib. Comput. **80**, 03–15 (2013)
6. P. Barnaghi, A. Sheth, C. Henson, From data to actionable knowledge: big data challenges in the web of things. IEEE Intell. Syst. **28**, 06–11 (2013)
7. R. Buyya, K. Ramamohanarao, C. Leckie et al., Big data analytics-enhanced cloud computing: challenges, architectural elements, and future direction. IEEE (2015). https://doi.org/10.1109/ICPADS.2015.18
8. A. Cardenas, P.K. Manadhata, S.P. Rajan, Big data analytics for security. IEEE Secur. Priv. **11**(6), 74–76 (2013)
9. H. Chen, R.H.L. Chiang, V.C. Storey et al., Business intelligence and analytics: from big data to big impact. MIS Q **36**(4), 1165–1188 (2012)
10. M. Chen, S. Mao, Y. Liu, Big data: a survey. Mobile Netw. Appl. **19**, 171–209 (2014)
11. J. Chen, Y. Chen, X. Du et al., Big Data challenge: a data management perspective. Environ. Front. Comput. Sci. **7**(2), 157–164 (2013)
12. P. Chen, C.Y. Zhang, Data-intensive applications, challenges, techniques and technologies: a survey on Big Data. Inform. Sci. **275**, 314–347 (2014)
13. T. Davenport, D.J. Patil, Data scientist: the sexiest job of the 21st century. Harv. Bus. Rev. (2012). https://doi.org/10.1080/01639374.2016.1245231
14. Dawar N (2016) Use Big Data to create value for customers, not just target them. Harv. Bus. Rev. https://hbr.org/2016/08/use-big-data-to-create-value-for-customers-not-just-target-them. Accessed 8 Sept 2017
15. S. Giest, Big data for policymaking: fad or fasttrack? Policy Sci. **50**(3), 367–382 (2017)
16. A. Gandomi, M. Raider, Beyond the hype: big data concepts, methods, and analytics. IJIM **35**(2), 137–144 (2015)
17. S. Earley, The digital transformation: staying competitive. IT Prof. **16**, 58–60 (2014)
18. S. Earley, Analytics, machine learning, and the internet of things. IT Prof. **17**, 10–13 (2015)
19. N. Elgendy, A. Elragal, Big data analytics in support of the decision making process. Procedia Comput. Sci. **100**, 1071–1084 (2016)
20. K. Evans, Where in the world is my information? Giving people access to their data. IEEE Secur. Priv. **12**(5), 78–81 (2014)
21. C. Everett, Big data—the future of cyber-security or its latest threat? Comput. Fraud Secur. **9**, 14–17 (2015)
22. A. Gandomi, M. Haider, Beyond the hype: big data concepts, methods, and analytics. Int. J. Inform. Manage. **35**(2), 137–144 (2015)
23. B. Gupta, M. Goul, B. Dinter, Business intelligence and big data in higher education: status of a multi-year model curriculum development effort for business school undergraduates, MS graduates, and MBAs. Commun. Assoc. Inf. Syst. **36**, 450–476 (2015)
24. K. Gillon, S. Aral, C.Y. Lin et al., Business analytics: radical shift or incremental change? Commun. Assoc. Inf. Syst. **34**(1), 287–296 (2014)

25. D. Hand, Statistics and computing: the genesis of data science. Stat. Comput. **25**, 705–711 (2015)
26. P. Helland, If you have too much data, then 'Good Enough' is Good Enough. Commun. ACM **54**(6), 40–47 (2011)
27. C. Ishikiriyama, Big data: um panorama global através de análise da literatura e survey. Dissertation, Universidade Federal Fluminense (2016)
28. N. Kabir, E. Carayannis, Big data, tacit knowledge and organizational competitiveness. J. Intell. Stud. Bus. **3**(3), 220–228 (2013)
29. A. Katal, M. Wazid, R.H. Goudar, Big data: issues, challenges, tools and good practices. Contemp. Comput. (2013). https://doi.org/10.1109/ic3.2013.6612229
30. N. Khan, I. Yaqoob, I. Abaker et al., Big data: survey, technologies, opportunities, and challenges. Sci. World J. (2014). https://doi.org/10.1155/2014/712826
31. G.H. Kim, S. Trimi, J.H. Chung, Big-data applications in the government sector. Commun. ACM **57**, 78–85 (2014)
32. H. Koscielniak, A. Puto, Big Data in decision making process of enterprises. Procedia Comput. Sci. **65**, 1052–1058 (2015)
33. T. Kraska, Finding the needle in the big data systems haystack. IEEE Internet Comput. **17**(1), 84–86 (2013)
34. N. Kshetri, Big data's impact on privacy, security and consumer welfare. Telecommun. Policy **38**(11), 1134–1145 (2014)
35. D. Laney, 3-D data management: controlling data volume velocity and variety (2001), http://blogs.gartner.com/doug-laney/files/2012/01/ad949-3D-Data-Management-Controlling-Data-Volume-Velocity-and-Variety.pdf. Accessed 20 Dec 2015
36. D. Laney, Gartner predicts three big data trends for business intelligence (2015), http://www.forbes.com/sites/gartnergroup/2015/02/12/gartner-predicts-three-big-data-trends-for-business-intelligence/#5cc6fd8366a2. Accessed 20 Dec 2015
37. P.S.H. Leeflang, P.C. Verhoef, P. Dahlström et al., Challenges and solutions for marketing in a digital era. Eur. Manage. J. **32**(1), 01–12 (2014)
38. J. Lin, Is big data a transient problem? IEEE Internet Comput. **16**(5), 86–90 (2015)
39. S. Liu, W. Cui, Y. Wu et al., A survey on information visualization: recent advances and challenges. Vis. Comput. **30**(12), 1373–1393 (2014)
40. M. Maciejewski, To do more, better, faster and more cheaply: using big data in public administration. Int. Rev. Adm. Sci. **83**(1S), 120–135 (2017)
41. T. Matzner, Why privacy is not enough privacy in the context of and big data. J. Inf. **12**(2), 93–106 (2014)
42. V. Mayer-Schonberger, K. Cukier, Big Data: como extrair volume, variedade, velocidade e valor da avalanche de informação cotidiana. Elsevier (2013)
43. D.E. O'Leary, Artificial intelligence and big data. IEEE Intell. Syst. **28**, 96–99 (2013)
44. D.J. Power, Using 'Big Data' for analytics and decision support. J. Decis. Syst. **23**(2), 222–228 (2014)
45. A. Picciano, The evolution of big data and learning analytics in American higher education. J. Asynchronous Learn. Netw. **16**(3), 09–21 (2012)
46. F. Provost, T. Fawcett, *Data Science and Its Relationship to Big Data and Data-Driven Decision Making* (Mary Ann Liebert, Inc., 2013). https://doi.org/10.1089/big.2013.1508
47. J. Reyes, The skinny on big data in education: learning analytics simplified. Techtrends **59**(2), 75–80 (2015)
48. M. Scherman, H. Krcmar, H. Hemsen et al., Big Data: an interdisciplinary opportunity for information systems research. Bus. Inf. Syst. Eng. **6**(5), 261–266 (2014)
49. J.P. Shim, A.M. French, J. Jablonski, Big data and analytics: issues, solutions, and ROI. Commun. Assoc. Inf. Syst. **37**, 797–810 (2015)
50. P. Tambe, Big data investment, skills, and firm value. Manage. Sci. **60**(6), 1452–1469 (2014)
51. J. Tien, Big Data: unleashing information. J. Syst. Sci. Syst. Eng. **22**(2), 127–151 (2013)
52. W.M. To, L. Lai, Data analytics in China: trends, issues, and challenges. IT Prof. **17**(4), 49–55 (2015)

53. S. Vlaene, Data scientists aren't domain experts. IT Prof. **15**(6), 12–17 (2013)
54. Z. Xiang, J.H. Gerdes, Z. Schwartz et al., What can big data and text analytics tell us about hotel guest experience and satisfaction? Int. J. Hospitality Manage. **44**, 120–130 (2015)
55. Y. Xiao, L.Y.Y. Lu, J.S. Liu et al., Knowledge diffusion path analysis of data quality literature: a main path analysis. J. Informetr. **8**(3), 594–605 (2014)
56. S.F. Wamba, S. Akter, A. Edwards et al., How 'big data' can make big impact: findings from a systematic review and a longitudinal case study. Int. J. Prod. Econ. **165**, 234–246 (2015)
57. M. Waller, S. Fawcett, Data science, predictive analytics, and big data: a revolution that will transform supply chain design and management. J. Bus. Logistics **34**(2), 77–84 (2013)
58. H. Watson, Tutorial: big data analytics: concepts, technologies, and applications. Commun. Assoc. Inf. Syst. **34**(65), 1247–1268 (2014)
59. B. Wixom, T. Ariyachandra, D. Douglas et al., The current state of business intelligence in academia: the arrival of big data. Commun. Assoc. Inf. Syst. **34**, 01–13 (2014)
60. X. Wu, X. Zhu, G.Q. Wu et al., Data mining with big data. IEEE Trans. Knowl. Data Eng. **26**(1), 97–107 (2014)

Celia Satiko Ishikiriyama Graduated in Statistics and Post Graduate in Knowledge Management. In 2017, accomplished a Master's degree in Production Engineering from Universidade Federal Fluminense. During the Master's, published papers in both national and international journals and now consults as a peer reviewer for Business Intelligent papers. Passionate about data analysis and a proven solid professional history in Business Intelligence spanning almost a decade in different market sectors. For the past two years a new professional challenge moved the focus to Demand Planning and Forecasting. Currently living in New Zealand and working with Demand Planning in a leading New Zealand wine company.

Carlos Francisco Simoes Gomes Doctor and Master of Production Engineering, Post-Doctorate in Mathematics. Project Manager and Researcher in Brazilian Naval System Center (CASNAV) from 1997 to 2007. After 2007, acted as Vice-Director in this same research center up to 2008. From 2006 until 2012, was Vice-President of Brazilian Operational Research Society. Has published 89 papers in both national and international journals. Supervised 37 graduating students, 24 master degrees students, 3 post-degree students and 1 doctorate thesis, author of 4 published books. Has taken part in almost one hundred congresses and symposiums in Brazil and abroad. Was part of the Brazilian delegation with the International Maritime Organization in the negotiations of the *Ballast Water and Double Hull Agreement*, as the author of the work that became the Brazilian position. Currently acts as Adjunct Professor of the Universidade Federal Fluminense, in addition to being a technical reviser for 20 different journals.

Chapter 4
Big Data for Predictive Analytics in High Acuity Health Settings

John Zaleski

Abstract Automated data capture is more prevalent than ever in healthcare today. Electronic health record systems (EHRs) and real-time data from medical devices and laboratory equipment, imaging, and patient demographics have greatly increased the capability to closely monitor, diagnose, and administer therapies to patients. This chapter focuses on the use of data for in-patient care management in high-acuity spaces, such as operating rooms (ORs), intensive care units (ICUs) and emergency departments (EDs). In addition, a discussion of various types of mathematical techniques and approaches for identifying patients at risk will be discussed as well as the identification and challenges associated with issuing of alarm signals on monitored patients.

Keywords Real-time data · Alarm signal · Medical device data
Adverse event · Monitoring · Wavelet transforms · Kalman Filter
Vital signs · Early warning · Periodograms

4.1 Patient Safety and Public Policy Implications

In 2013, The Joint Commission (TJC) issued a Sentinel Event Alert (SEA) in which it called for a clarion call for action nationally regarding the issue of medical device alarm safety in hospitals [1]. The Joint Commission referenced the 2011 Medical Device Alarm Summit convened by AAMI [2] which reported that "between 85 and 99% of alarm signals do not require clinical intervention..." principally because alarms issued by bedside physiologic monitors, pulse oximeters, and monitors of blood pressure as well as infusion pumps and mechanical ventilators issue an overwhelming quantity of alarm signals that are non-actionable clinically due to artifact, dried out sensor pads, or poorly-positioned sensors.

J. Zaleski (✉)
Bernoulli, Enterprise, Inc., Milford, CT, USA
e-mail: jzaleski@bernoullihealth.com
URL: http://bernoullihealth.com

A. Emrouznejad and V. Charles (eds.), *Big Data for the Greater Good*,
Studies in Big Data 42, https://doi.org/10.1007/978-3-319-93061-9_4

The Joint Commission estimated from their Sentinel Event database that the resulting alarm fatigue associated with receiving so many non-actionable alarm signals resulted in a true patient safety concern as it was difficult to discern between truly actionable and false alarm signals. Estimates of events between January 2009 and June 2012 revealed as many as 98 alarm-related events, and of this quantity, 80 resulted in death.

The most common injuries or deaths pertaining to alarm signals included patient falls, delays in treatment, mechanical ventilator-related and medication errors which were directly related back to alarm signal issues. The findings further revealed that the majority of these alarms occurred within the inpatient hospital setting, mostly involving the intensive care unit, general care floor, and emergency department settings. Specific causes included absent or inadequate alarm systems, inappropriate alarm signal settings, inaudible alarms, and alarms being turned off. In 2017, The ECRI reported [3] that item number three (3) in their top 10 health technology hazards for 2017 pertained to missed ventilator alarms leading to patient harm.

The Joint Commission issued a separate report on alarm system safety identifying national patient safety goal *NPSG.06.01.01: Improve the safety of clinical alarm systems* [4]. Priorities highlighted from this "*Requirement, Rationale, Reference*" report include:

- EP 1: "As of July 1, 2014…establish alarm system safety as a hospital priority," and
- EP 2: "During 2014, identify the most important alarm signals…"

That alarm signal safety is a recognized issue was supported based on a survey issued by the Joint Commission in which 1600 responses were received and, of these, "90% of hospital respondents agreed…[but]… 70% believed alarms were effectively managed…[and] fewer than 50%…had an organization-wide process for alarm management."

Management of alarm signal traffic inside the hospital requires effective measurement of the source data. The old adage that one cannot control what one cannot measure applies here. In order to step through this process, a treatment of the state of data capture will start off the discussion, followed by several examples and implications for establishing a process for effective management and oversight of alarm systems and their related signals.

4.2 Introduction: The State of Data Capture in Healthcare

In this section the types of data that are typically collected on patients is discussed, comprising both automated and manually recorded observations and findings. The challenges associated with automated data collection from patient care devices in terms of proprietary queries and translation of data are highlighted, and the semantic interpretation of data is discussed.

4.2.1 Patient Data and Demographics

Modern medicine employs data from multiple sources to guide treatment therapies for, research, and predict the outcomes in patients. Many sources of data are available, and the primary sources include manual observations by nursing and physicians and other allied health professionals, imaging data from magnetic resonance imaging (MRI), X-Ray, computed tomography (CT), ultrasound (US), positron emission tomography (PET), and time-based signals such as issued from electrocardiograms and other point of care medical devices, including multi-parameter physiologic monitors, mechanical ventilators, anesthesia machines, etc.

Big data, characterized by Gartner as "high-volume, high-velocity, high-variety" are often maintained in silos in healthcare organizations [5, 6]. The data used for diagnosis, treatment and therapy is managed in many places within the healthcare system. The principal location is the EHR. EHRs in their current state have limited capacities for certain types of data, such as high-frequency and high-bandwidth information like physiologic monitoring waveforms and real-time alarm signals received from medical devices for the purposes of guiding intervention.

One way to depict the relationship between the patient and data obtained from the patient is through the diagram of Fig. 4.1. Information regarding the state of the patient, as pertains to this figure, are derived from patient oral histories, clinical observations (i.e.: subjective sources) and from machine-measured data (i.e.: objective sources). This information contains fact-based as well as fuzzy information. For example, certain types of subjective questions can result in answers that are fuzzy or approximate:

- "how often have you felt pain?"
- "when did you first notice the problem?"
- "how many times a week does it occur?"
- "did your parents have similar ailments?"

While subjective information may be approximate or imprecise, objective information, such as that derived from medical devices, is "the least error-prone," [7] as this class of data is typically free of artifact and subjective interpretation.

The types of data, or findings, associated with the patient typically fall into categories of:

(a) Imagery;
(b) Vital signs and other time-based signals; and,
(c) Clinical observations.

Observations and findings are maintained in the EHR—a repository which maintains the patient historical and current clinical record, observations, orders and clinician-entered observations. Time-based signals are those produced by physiologic monitors, mechanical ventilators, anesthesia machines, spot vital signs monitors and related equipment. Data consumed by the clinician are exemplified in the diagram of Fig. 4.2. Here, medical device data (including vital signs and imagery),

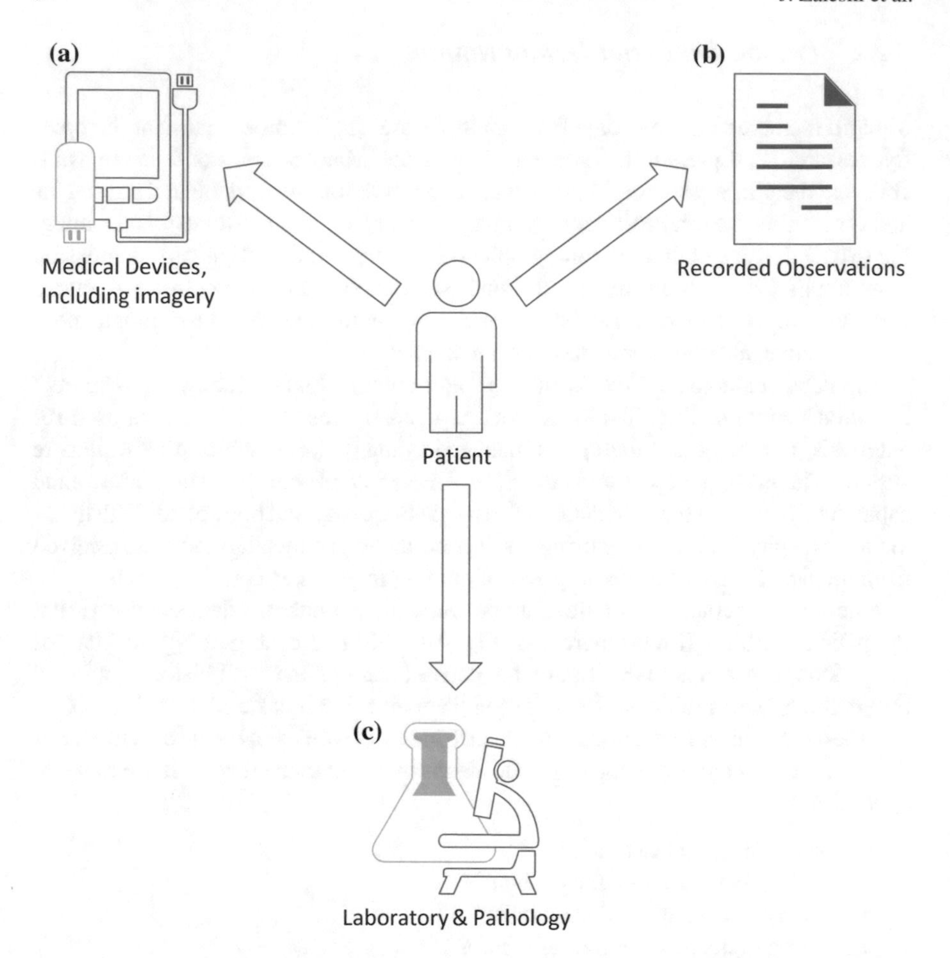

Fig. 4.1 Examples of types of data obtained from the patient. The three broad categories include **a** medical devices, such as cardiac monitors, mechanical ventilators, X-Ray machines, Magnetic Resonance Imaging machines, and the like; **b** recorded observations from clinicians; and, **c** laboratory measurements, such as complete blood counts (CBC) or arterial blood gas (ABG), and pathology reports

patient history (historical records), population data associated with patients of similar demographics, diagnoses, and outcomes, laboratory and pathology reports, and research in the form of publications and studies, are all provided as input to the clinical end user. These separate sources of information usually are in different formats and are translated into actionable information within the mind of the clinician. These sources also arrive at differing rates. Some sources update data frequently (e.g.: every second) while others are not updated at all (e.g.: research publications) and represent static sources of knowledge.

Data capture in healthcare information technology (HIT) systems falls into the categories of:

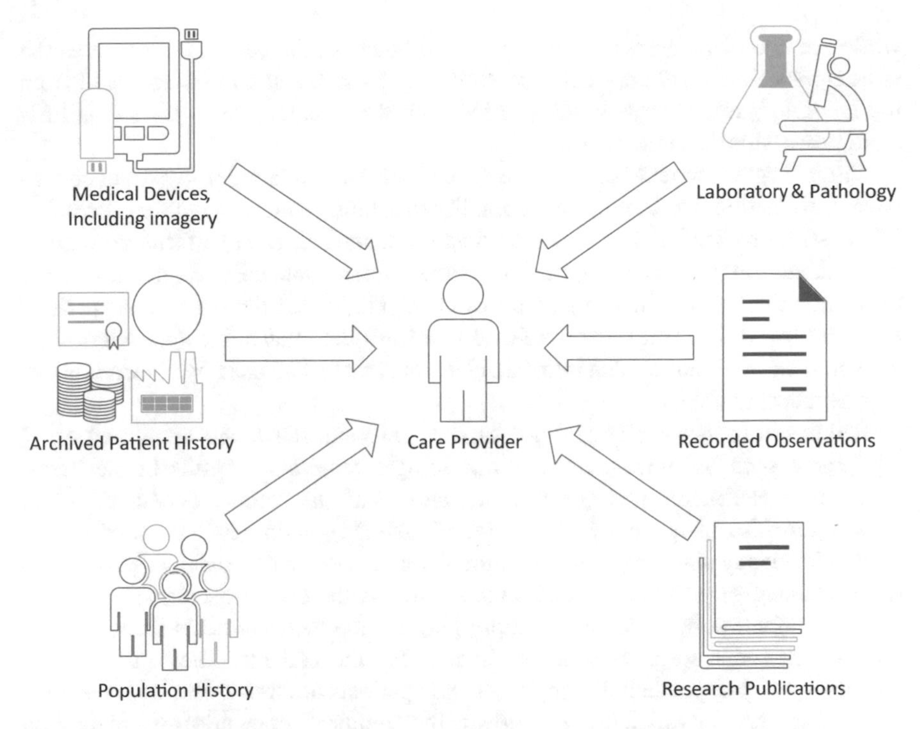

Fig. 4.2 Examples of data types used by clinicians to diagnose, treat and manage patients

- Discrete data capture—measurements collected directly from medical devices through discrete data collection mechanisms.
- Real-time data capture—real-time data collected at high frequency from physiologic or multi-parameter monitors or ancillary devices.
- Alarm signals—emergent annunciations of signals from medical devices indicating potentially clinically-relevant events.
- Imagery—medical images from magnetic resonance imaging (MRI), X-rays, computed tomography (CT), ultrasound (US), and positron emission tomography (PET) scans.
- Clinical notes—subjective observations in free text from EHR nursing and physician entries.
- Laboratory data—results of laboratory panels based upon blood or specimen samples taken from patients.

Discrete data collection is normally available from multi-parameter physiologic monitors (e.g.: vital signs time series-based capture), mechanical ventilators, and anesthesia machines. Higher frequency and real-time data, such as waveforms and alarm signals, can be collected in ranges from sub-second to multi-second intervals. These data can be automatically collected using middleware or from the medical devices themselves. Imagery data can be collected from the equipment automatically while the rate of imagery capture is often ad hoc and, thus, not as frequent as the vital signs data. Laboratory data are normally discrete and can be collected from the laboratory information systems (LIS). Yet, these data are also ad hoc and not continuous, similar to imagery data.

Clinical notes are text-based and require interpretation in order to extract quantitative information, which may be accomplished through natural language processing (NLP) software. Patient demographics, diagnostic and contextual information can be determined from the free text contained within clinical notes. Hence, it is important to include this information to gain a full understanding of the state of the patient. Any text-based data must be discretized in a form that makes the data searchable, interpretable, and translatable into data elements that can be used by algorithms and mathematical methods.

Patients in higher acuity settings, such as ORs and ICUs, are monitored using equipment such as multi-parameter physiologic monitors, anesthesia machines, mechanical ventilators and specialty monitors such as cardiac output, end-tidal carbon dioxide, bi-spectral index, cerebral oximetry monitors, and others. The higher-frequency data normally monitored on higher acuity patients pertains to life-sustaining or interventional functions, such as the cardiovascular, respiratory, and endocrine systems. In higher acuity patients the measurements from the cardiovascular and respiratory systems "speak" for the patients. These patients are technologically-dependent. Examples of such patients are those who have received invasive surgery for which post-operatively they require therapeutic instrumentation to maintain their life systems (life support equipment). Mechanical ventilation is an example of a life-sustaining piece of equipment, and patients in whom respiratory and/or cardiovascular systems have been compromised, or who have experienced trauma, such as traumatic brain injury, such equipment is necessary to sustain life. Hence, critical changes in parameter measurements over time, such as unanticipated changes in cardiac output, breathing patterns, blood oxygenation levels, and cerebral pressures, is essential to ensuring interventions can be performed prior to the occurrence of a catastrophic event (i.e.: death).

Critical changes in measurements or deviations from normal or baseline values are reflected as alarm signals, often transmitted to clinical staff. Such alarm signals are issued when a monitored value breaches a clinically significant boundary or threshold value. Boundary breaches may not necessarily be detected through individual ad hoc measurements. Indeed, critical changes in patient values may only be visible or detected through continuous monitoring. Continuous monitoring (i.e.: higher frequency monitoring, including at waveform levels) is necessary to ensure capturing and identifying local extrema (maximum and minimum values) in the data signal. Discrete monitoring at lower frequencies often misses these local extrema.

Consider Fig. 4.3, which shows a notional time series of data corresponding to a parameter obtained from a point-of-care medical device.

When considering discretized data collection from medical devices, the frequency of data collection needs to be at a rate whereby important changes or values can be detected. In this figure, if receiving systems (such as the EHRs) cannot receive the

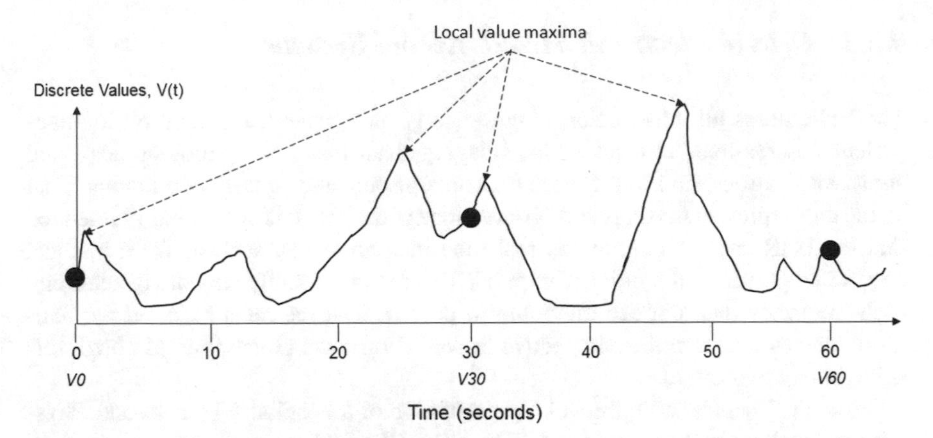

Fig. 4.3 Illustration of continuous signal monitoring of a parameter value, $V(t)$, with discrete samples at 0, 30 and 60 s. The identification of local value maxima are not detected in the discrete data

continuous data collection, then, not all data will be received from the medical device but, rather, a subset corresponding to the data collection rate requirements of the EHR.

For example, if the EHR can receive data at a rate no greater than once per minute, but data can be transmitted from the fastest medical device at the rate of once per second, then only one sample per minute from that medical device can be posted to the EHR.

Various policies may be employed in the selection of the sample from the medical device: last measurement, first measurement, median measurement in a given time window, or other. For example:

- Taking measurements at the start of an interval (in this case, the start of a 1 min interval);
- Taking measurements at the end of an interval (i.e.: the end of the 1 min interval in this case);
- Taking measurements at various locations in between; or,
- Collecting measurements continuously and processing them as they are received in real-time.

There may be local minima and maxima that exceed clinical thresholds during this interval which will not be reported, depending on the data sampling interval and the ability of the receiving system to accept data at the highest rate possible. The point is that the receiving system must be able to accept the data at the required rate to meet the clinical need, whether the data are discrete or continuous.

4.2.2 Data in Electronic Health Record Systems

The EHR stores all information longitudinally on a patient. Information includes patient observations, clinical notes, imaging data, histories, demographics, vital signs, medications, findings. Stored data are episodic and "quasi" continuous. That is, the data represent findings and observations over time. It is not normally intended that the EHR capture continuous real-time measurements, such as from medical devices employed at the point of care: EHRs are not normally capable of retaining high-frequency data nor are they able to process it at the rates required for real-time intervention from the perspective of waveform-level processing (although this is beginning to change).

Figure 4.3 depicts a high-level representation of a hospital HIT network. Physiologic monitors that are used in ICUs, EDs, ORs and other high acuity settings monitor the patient's vital signs, including cardiac function. Devices for respiratory function and maintenance, such as mechanical ventilators and anesthesia machines, are therapeutic devices used to deliver oxygen and anesthetic gases to the patient. Infusion pumps deliver intravenous drugs to the patient. Physiologic monitors that communicate data as part of a central monitoring network, CMN (i.e.: a closed network capable of providing real-time capture and display of individual waveforms) can also deliver data to the EHR system via networking protocols such as Transmission Control Protocol (TCP) or universal datagram protocol (UDP) and Internet Protocol (IP) (e.g.: TCP/IP) over Ethernet.

Some medical devices can only communicate data via serial data communication using RS 232 physical interconnections and a data translation bridge in the form of a Serial-to-Ethernet data converter. The function of the Serial-to-Ethernet data converter together with subsequent downstream processing is illustrated in Fig. 4.4. Medical devices often communicate data through a proprietary query-response mechanism. The translation from proprietary data to the more standardized formats of Health Level Seven (HL7) messaging over Ethernet using TCP/IP is required for communication with EHRs. The high-level process for data collection and communication is described in the flow diagram of Fig. 4.4. The outbound data are normally translated from machine-proprietary-specific to that of a standardized format. An example of such a data translation from proprietary to the healthcare standard HL7 is depicted in Tables 4.1 and 4.2, respectively. Table 4.1 shows the raw data received from a Puritan Bennett 840 mechanical ventilator. The data elements contain machine settings and patient measurements. Table 4.2 shows these data translated into an HL7 unsolicited result message. The data messaging standard HL7 is flexible and newer forms of the standard (e.g.: HL7 Fast Healthcare Interoperability Resources, FHIR®) are evolving to facilitate improved data communication to EHRs.

Examples of Serial-to-Ethernet data collection appliances abound in the industry and can be found through commercial vendors such as Bernoulli Enterprise, Inc. [8] and Qualcomm [9] (Fig. 4.5).

The data presented in the above table are in accordance with HL7 version 2.3 standard. In the case of the Puritan Bennett 840 mechanical ventilator, the data are

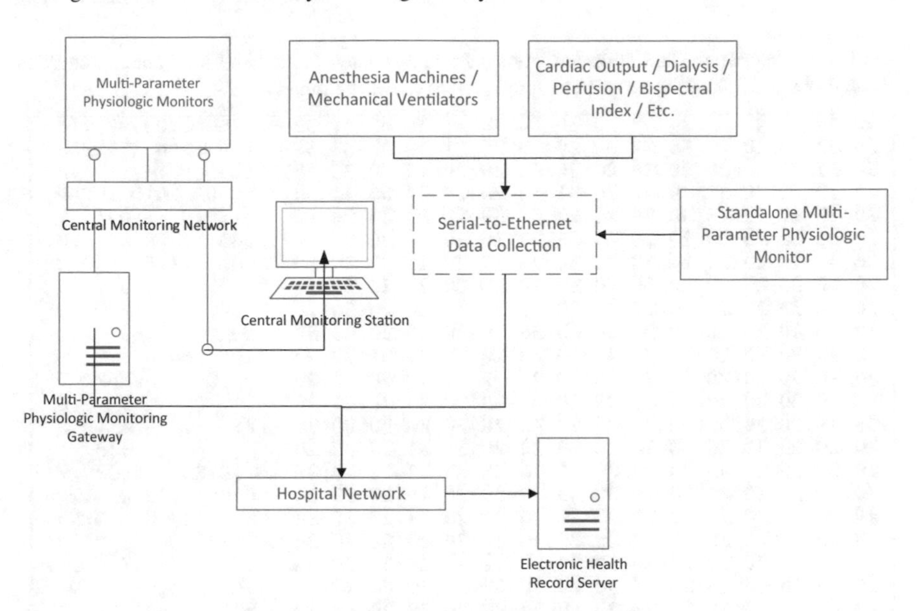

Fig. 4.4 Network diagram depicting connectivity among medical devices and EHR system within the HIT network. Physiologic monitors used in most high acuity settings such as critical care and operating rooms communicate their data via a central monitoring network (CMN), which can feed data through a Physiologic Monitoring Gateway that effectively isolates the higher frequency waveforms from the hospital network. Data from physiologic monitors connected in this way are also visible as waveforms on central monitoring stations (CMS) which enable clinical viewing of alarms, waveforms and events on each patient. Discrete data collection as well as some waveform data can be drawn using Serial-to-Ethernet bridges, whereby data from certain types of high acuity equipment, including mechanical ventilators, anesthesia machines and standalone multi-parameter physiologic monitors can be communicated to the EHR over the hospital network

in American Standard Code for Information Interchange (ASCII) format. There can be great variety in terms of the formats and features of the data delivered by medical devices: the type, format, and semantic naming of the data they issue and the frequency with which they can issue data. This variety and the proprietary nature of many medical devices make it necessary to transform data into common formats and messaging representations so as to ensure consistent interpretation since some medical devices which perform the same function may employ different semantics. To this end, several initiatives have focused on aligning terminology and semantic representations from data derived from medical devices. Among these are the Rosetta Terminology Mapping Management Service (RTMMS) which, in partnership with the National Institute for Standards and Technology (NIST) has adopted and is refining a harmonized set of standards on unified codes, units of measure, and normative mappings [10].

Table 4.1 Proprietary data collected from a Puritan-Bennett model 840 mechanical ventilator through RS 232 port. Settings and measured patient values are present

```
4D 49 53 43 41 2C 37 30 36 2C 39 37 2C 02 31 33   MISCA,706,97,.13
3A 32 36 20 2C 38 34 30 20 33 35 31 30 30 38 33   :26 ,840 3510083
36 37 35 20 20 20 20 2C 20 20 20 20 20 20 2C 53   675    ,      ,S
45 50 20 30 33 20 32 30 31 30 20 2C 43 50 41 50   EP 03 2010 ,CPAP
20 20 2C 30 2E 30 20 20 20 2C 30 2E 34 34 20 20     ,0.0   ,0.44
2C 36 35 20 20 20 20 2C 32 31 20 20 20 20 2C 30   ,65    ,21    ,0
2E 30 20 20 20 2C 33 2E 30 20 20 20 2C 30 2E 30   .0   ,3.0   ,0.0
20 20 20 2C 20 20 20 20 20 20 2C 20 20 20 20 20      ,      ,
20 2C 20 20 20 20 20 20 2C 20 20 20 20 20 20 2C    ,      ,      ,
32 32 20 20 20 20 2C 30 2E 36 30 20 20 2C 31 32   22    ,0.60  ,12
2E 30 20 20 2C 36 30 20 20 20 20 2C 32 31 20 20   .0  ,60    ,21
20 20 2C 30 20 20 20 20 20 2C 52 41 4D 50 20 20     ,0     ,RAMP
2C 20 20 20 20 20 20 2C 20 20 20 20 20 20 2C 4F   ,      ,      ,O
46 46 20 20 20 2C 20 20 20 20 20 20 2C 20 20 20   FF   ,      ,
20 20 20 2C 20 20 20 20 20 20 2C 31 32 20 20 20      ,      ,12
20 2C 30 2E 35 33 20 20 2C 36 2E 33 33 20 20 2C    ,0.53  ,6.33  ,
30 2E 30 20 20 20 2C 32 35 2E 30 20 20 2C 37 2E   0.0   ,25.0  ,7.
39 20 20 20 2C 32 32 2E 30 20 20 2C 33 2E 35 30   9   ,22.0  ,3.50
20 20 2C 35 30 20 20 20 20 2C 20 20 20 20 20 20     ,50    ,
2C 20 20 20 20 20 20 2C 30 2E 32 30 20 20 2C 31   ,      ,0.20  ,1
2E 30 20 20 20 2C 34 30 20 20 20 20 2C 4E 4F 52   .0   ,40    ,NOR
4D 41 4C 2C 20 20 20 20 20 20 2C 20 20 20 20 20   MAL,      ,
20 2C 4E 4F 52 4D 41 4C 2C 4E 4F 52 4D 41 4C 2C    ,NORMAL,NORMAL,
4E 4F 52 4D 41 4C 2C 41 4C 41 52 4D 20 2C 4E 4F   NORMAL,ALARM ,NO
52 4D 41 4C 2C 4E 4F 52 4D 41 4C 2C 41 4C 41 52   RMAL,NORMAL,ALAR
4D 20 2C 20 20 20 20 20 20 2C 20 20 20 20 20 20   M ,      ,
2C 31 33 3A 32 36 20 2C 20 20 20 20 20 20 2C 53   ,13:26 ,      ,S
45 50 20 30 33 20 32 30 31 30 20 2C 30 2E 30 20   EP 03 2010 ,0.0
20 20 2C 30 2E 30 20 20 20 2C 33 36 2E 30 20 20     ,0.0   ,36.0
2C 31 37 2E 30 30 20 2C 30 20 20 20 20 20 2C 30   ,17.00 ,0     ,0
2E 30 30 30 20 2C 30 20 20 20 20 20 2C 34 20 20   .000 ,0     ,4
20 20 20 2C 33 20 20 20 20 20 2C 20 20 20 20 20      ,3     ,
20 2C 20 20 20 20 20 20 2C 20 20 20 20 20 20 2C    ,      ,      ,
20 20 20 20 20 20 2C 20 20 20 20 20 20 2C 20 20         ,      ,
20 20 20 20 2C 20 20 20 20 20 20 2C 20 20 20 20       ,      ,
20 20 2C 20 20 20 20 20 20 2C 20 20 20 20 20 20     ,      ,
2C 20 20 20 20 20 20 2C 4E 4F 4F 50 20 43 2C 32   ,      ,NOOP C,2
32 2E 30 30 20 2C 30 20 20 20 20 20 2C 30 2E 30   2.00 ,0     ,0.0
30 20 20 2C 32 32 20 20 20 20 2C 30 20 20 20 20   0  ,22    ,0
20 2C 31 32 2E 30 20 20 2C 30 2E 30 30 20 20 2C    ,12.0  ,0.00  ,
32 31 20 20 20 20 2C 35 30 20 20 20 20 2C 4F 46   21    ,50    ,OF
46 20 20 20 2C 41 4C 41 52 4D 20 2C 4E 4F 52 4D   F   ,ALARM ,NORM
41 4C 2C 30 2E 30 30 20 20 2C 30 2E 30 30 20 20   AL,0.00  ,0.00
2C 30 2E 30 30 20 20 2C 30 2E 30 30 20 20 2C 20   ,0.00  ,0.00  ,
20 20 20 20 20 20 20 20 2C 31 3A 33 2E 35 30 2C           ,1:3.50,
03 0D
```

4.2.3 Available Imaging and Medical Device Data, by Department and by Patient Acuity

Medical devices are a key source of data in health care environments, and the vast majority of data collection from medical devices occurs in the ORs and ICUs [11, 12]. Patients in these locations are highly technologically-dependent and their data repre-

Table 4.2 Puritan Bennett model 840 mechanical ventilator data translated into HL7 messaging format

```
MSH|^~\&|source|dest.|||20160903133103||ORU^R01|IDC00501801|P|2.3
PID|12345678||||
PV1||I|
OBR|1||||||20160903133103||||||||IDC
OBX|1|ST|DEV-TIME^Ventilator time||13:27|
OBX|2|ST|DEV-ID^Ventilator ID||840 3510083675|
OBX|3|ST|DEV-DATE^Ventilator Date||SEP 03 2016|
OBX|4|TS|DEV-TS^Device Timestamp||20100309132700|
OBX|5|ST|VNT-MODE^Ventilator Mode||BILEVL|
OBX|6|NM|SET-RR^Respiratory rate||12.0|/min
OBX|7|NM|SET-TV^Tidal volume||0.00|L
OBX|8|NM|SET-MFLW^Peak flow setting||0|L/min
OBX|9|NM|SET-O2^O2% setting||21|%
OBX|10|NM|SET-PSENS^Pressure sensitivity||0.0|cmH2O
OBX|11|NM|SET-LO-PEEP^PEEP Low (in BILEVEL) setting||3.0|cmH2O
OBX|12|NM|IN-HLD^Plateau||0.0|cmH2O
OBX|13|NM|SET-APN-T^Apnea interval||22|s
OBX|14|NM|SET-APN-TV^Apnea tidal volume||0.60|cmH2O
OBX|15|NM|SET-APN-RR^Apnea respiratory rate||12.0|/min
OBX|16|NM|SET-APN-FLW^Apnea peak flow||60|L/min
OBX|17|NM|SET-APN-O2^Apnea O2%||21|%
OBX|18|NM|SET-PPS^Pressure support||0|cmH2O
OBX|19|ST|SET-FLW-PTRN^Flow pattern|||
OBX|20|ST|O2-IN^O2 Supply||OFF|
OBX|21|NM|VNT-RR^Total respiratory rate||12|/min
OBX|22|NM|TV^Exhaled tidal volume||0.33|L
OBX|23|NM|MV^Exhaled minute volume||5.64|L/min
OBX|24|NM|SPO-MV^Spontaneous minute volume||0.0|L
OBX|25|NM|SET-MCP^Maximum circuit pressure||20.0|cmH2O
OBX|26|NM|AWP^Mean airway pressure||7.2|cmH2O
OBX|27|NM|PIP^End inspiratory pressure||20.0|cmH2O
OBX|28|NM|IE-E^1/E component of I:E||5.80|
OBX|29|NM|SET-HI-PIP^High circuit pressure limit||50|cmH2O
OBX|30|NM|SET-LO-TV^Low exhaled tidal volume limit||0.20|L
OBX|31|NM|SET-LO-MV^Low exhaled minute volume limit||1.0|L
OBX|32|NM|SET-HI-RR^High respiratory rate limit||40|/min
OBX|33|ST|ALR-HI-PIP^High circuit pressure alarm status||NORMAL|
OBX|34|ST|ALR-LO-TV^Low exhaled tidal vol. alarm status||NORMAL|
OBX|35|ST|ALR-LO-MV^Low exhaled minute vol. alarm stat||NORMAL|
OBX|36|ST|ALR-HI-RR^High respiratory rate alarm status||NORMAL|
OBX|37|ST|ALR-NO-O2^No O2 supply alarm status||ALARM|
OBX|38|ST|ALR-NO-AIR^No air supply alarm status||NORMAL|
OBX|39|ST|ALR-APN^Apnea alarm status||RESET|
OBX|40|NM|SET-FLW-BASE^Ventilator-set base flow||4|L/min
OBX|41|NM|SET-FLW-TRG^Flow sensitivity setting||3|L/min
OBX|42|NM|PIP^End inspiratory pressure||20.00|cmH2O
OBX|43|NM|SET-PIP^Inspiratory press. or PEEP High set.||18|cmH2O
OBX|44|NM|SET-INSPT^Insp. time or PEEP High time set||0.74|s
OBX|45|NM|SET-APN-T^Apnea interval setting||22|s
OBX|46|NM|SET-APN-IP^Apnea inspiratory pressure setting||0|cmH2O
OBX|47|NM|SET-APN-RR^Apnea respiratory rate setting||12.0|/min
OBX|48|NM|SET-APN-IT^Apnea inspiratory time setting||0.00|s
OBX|49|NM|SET-APN-O2^Apnea O2% setting||21|%
OBX|50|NM|SET-PMAX^High circuit pressure limit||50|cmH2O
OBX|51|ST|ALR-MUTE^Alarm silence state||OFF|
OBX|52|ST|ALR-APN^Apnea alarm status||RESET|
OBX|53|ST|ALR-VNT^Severe Occl./Disconn. alarm status||NORMAL|
OBX|54|NM|SET-HL-HI^High comp. of H:L (Bi-Level) setting||1.00|
OBX|55|NM|SET-HL-LO^Low comp. of H:L (Bi-Level) setting||5.76|
OBX|56|ST|SET-APN-IEI^Inspiratory comp. apnea I:E ratio||0.00|
OBX|57|ST|SET-APN-IEE^Expiratory comp. of apnea I:E ratio||0.00|
OBX|58|ST|SET-CONST^Const rate set. chn./ PCV mand. brths||I-TIME|
OBX|59|ST|IE^Monitored value of I:E ratio||1:5.80|
```

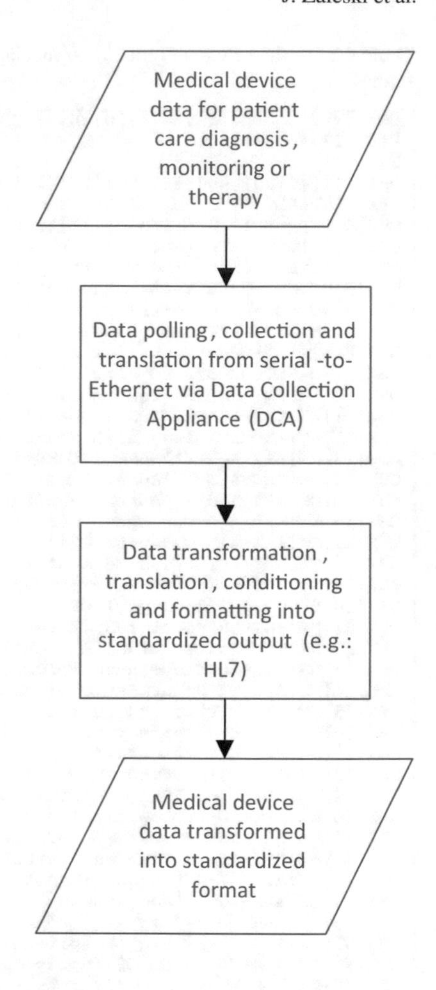

Fig. 4.5 Conversion from proprietary to standardized data messaging formats

sent the states of their cardiovascular and respiratory systems measured through continuous vital signs monitoring. Typically, ORs feature anesthesia machines, continuous vital signs monitoring obtained through physiologic monitors, cerebral oximetry monitors, cardiac output monitors, and cardiothoracic bypass perfusion machines. ICUs support continuous vital signs monitoring via physiologic monitors, mechanical ventilators and intra-aortic balloon pumps (IABPs), hemodialysis, pulse oximetry monitors, and infusion pumps. Medical-surgical wards or step-down units outside of ICUs often feature ad hoc spot vital signs monitoring, infusion pumps, and end tidal carbon dioxide monitoring. Emergency Departments (EDs) often feature continuous and spot vital signs monitoring and infusion pumps. Endoscopy, imaging (e.g.: Interventional Radiology (IR), MRI, CT, and catheterization labs (Cath Labs) can employ physiologic monitors and anesthesia machines, as well as continuous and spot vital signs monitoring.

Imaging data are obtained from X-Ray, MRI, CT, PET, and US. Images are taken as part of normal and interventional care. Due to the facilities, environmental and logistics requirements associated with MRI, CT and PET scans, patients are brought to the specific locations where these machines are housed for imaging. X-Ray can be portable and, particularly in ICUs and emergency departments (EDs), these machines can be brought to the patients.

4.2.4 *How Data Are Used to Diagnose, Treat and Manage the Patient*

Data are essential to clinical decision making. Yet, the data need to be studied, analyzed and put to use effectively. This requires intelligence in the form of algorithms that can be used to crunch and interpret the raw essence of patient data [13]. The patient comprises many subsystems (e.g.: respiratory, cardiovascular, endocrine, etc.) Hence, the patient can be considered as a system-of-systems, in which system-level modeling representing overall behavior of the patient as a function of inputs (e.g.: drug dosages) and outputs (e.g.: cardiovascular and respiratory behavior) considers the patient attempts to diagnose and predict or anticipate overall system behavior on the basis of the interaction of the various subsystems [14]. The following sections detail the use of data for prediction with examples of various techniques that can be applied for analytical purposes. These techniques are also useful in creating of alarm signals which can be used to aid intervention based on measured features of the various subsystems and how they meet or breach nominal threshold values during the course of treatment throughout the process of patient care management.

4.2.5 *Ways to Visualize Data in Healthcare Environments to Facilitate End-User (i.e., Clinical User) Consumption*

High acuity wards within healthcare systems employ central monitoring stations (CMS) where clinical staff monitor patients. In higher acuity settings such as ICUs, patient monitoring involves communicating measurements (e.g.: individual sensor measurements and EKG waveforms) over dedicated networks from each patient room to this CMS for display of waveforms, alarm signals and vital sign measurements. The CMS presents a multi-patient view of both real-time data and discrete measurements from the physiologic monitors at the bedside. Discrete data are often communicated from the CMN to the EHR.

Furthermore, the EHR system presents a portal (most often Web-based) through which the clinical end user can interact in terms of viewing patient state, entering observations and orders, recording drug administrations, intakes and outputs, and vital signs data as measured using the multi-parameter monitors. These two sources of visualization (CMS and EHR) provide the principal vehicles for clinical end user interaction. Typical methods for displaying data include computer monitors at nursing stations, tracking board monitors (i.e.: large liquid crystal display, LCD, panels) that are easily visible by clinical staff, and workstations on wheels that roll between patient rooms and with which clinical staff interact.

4.3 Predictive and Prescriptive Analytics Methods Used in Healthcare

This section focuses on examples of mathematical techniques as well as written algorithms used in the processing of real-time clinical data for the purpose of real-time clinical decision making. Key sources of data are medical devices used in patient care, which are a potential source for continuous data collection.

4.3.1 Protocols and Clinical Practice Guidelines

Discussion on the topic of prescriptive methods in healthcare would not be complete without a discussion of protocols and clinical practice guidelines (CPGs). Protocols and CPGs are, in effect, algorithms. The definition of CPGs was proffered by Field and Lohr in the early 1990s is as follows:

> Clinical Practice Guidelines are systematically developed statements to assist practitioners and patient decisions about appropriate healthcare for specific clinical circumstances. [15]

CPGs include standards, algorithms, rules that define practices associated with various aspects of care. The National Guidelines Clearinghouse, in their 2013 revised Criteria for Inclusion of CPGs, states that the CPG "…contains systematically developed statements…intended to optimize patient care…" [16]. In general, CPGs can represent algorithms, methods, rules and statements that provide guidance on clinical practice related to specific patient conditions or treatment protocols. In this sense CPGs represent algorithmic approaches for guiding specific types of care. But, as the name implies, they guide treatment protocols but do not restrict the clinician from deviating for a specific patient based on clinical judgment. In short, CPGs represent routes or pathways for care, but the clinician is free to contemplate care outside of prescribed protocols based on judgment. The specific types and methods described within CPGs is beyond the scope of this chapter. Yet, for the interested reader, a number of web sites provide insight into the many available CPGs. Examples of CPGs can be found at the following sites:

- http://www.Openclinical.org features CPGs related to acute care systems, decision support systems, educational systems, laboratory systems, medical imaging, and quality assurance and administration.
- The American College of Physicians (ACP), https://www.acponline.org/clinical-information/guidelines features CPGs, recommendations, educational information, and general resources for CPG development.

- AHRQ's national guideline clearinghouse is a public resource for summaries of evidence-based CPGs, located at http://www.guidelines.gov. Guidelines in the following areas are detailed:
  - Cardiology
  - Internal medicine
  - Physical medicine and rehabilitation
  - Critical Care
  - Nursing
  - Psychiatry
  - Emergency medicine
  - Obstetrics and gynecology
  - Family practice
  - Oncology
  - Surgery
  - Geriatrics
  - Orthopedic surgery
  - Infectious diseases
  - Pediatrics

4.3.2 Clinical Decision Making Using Multi-source and Multi-parameter Data

Clinical decision making relies upon data from multiple sources, including the aforementioned medical devices, laboratory findings, imaging data, and drug delivery systems (infusion and medication administration systems) and subjective sources (i.e.: non-automated) such as clinical notes and observations by care providers. Data from these sources are used for patient care management, charting patient condition, and to annunciate events that require intervention. Event annunciations are also known as alarm signals issued by vital signs monitoring and other diagnostic and therapeutic equipment, and are intended to notify a clinical end user when something has gone awry with the patient. In Sect. 4.4.5 a more descriptive discussion of alarms and alarm signal annunciation is undertaken. Multi-source, multi-parameter data can be used in conjunction with CPGs to provide assistance and guide clinical decision making. Data combined from several sources in the form of rules that can serve to notify clinical end users to provide indications of impending clinical conditions.

For example, notifying a clinical end user when certain measurements or findings reach specific levels can imply clinically significant events, such as the potential onset of blood-borne infections, respiratory depression, or cardiac arrest. Models that incorporate multivariate data into assessments abound. An example of a multivariate protocol is the Early Warning Score for Rapid Response Team (RRT) notification, provided in abbreviated form from the Institute for Clinical Systems Improvement (ICSI) [17, p. 13] and summarized in Table 4.3. Developing the rapid response score

Table 4.3 Early warning score protocol to initiate rapid response

Score	3	2	1	0	1	2	3
Heart rate (beats/minute)		<40	41–50	51–100	101–110	111–130	>130
Systolic blood pressure (mmHg)	<70	71–80	81–100	101–179	180–199	200–220	>220
Respiratory rate (breaths/minute)		<8	8–11	12–20	21–25	26–30	>30
4-h urine output (milliliters)	<80	80–120	120–200		>800		
Central nervous system			Confusion	Awake and responsive	Responds to verbal commands	Responds to painful stimuli	Unresponsive
Oxygen saturation (%)	<85%	86–89%					
Respiratory support /oxygen therapy	Bi-PaP/CPAP	Hi-flow	O_2 therapy				

From the Institute for clinical systems improvement. Initiate a rapid response if the sum of all individual scores is 3 or greater. Copyright 2011 by the Institute for clinical systems improvement. Used with permission. ICSI retains all rights and material will in no way be used to determine provider compensation

requires input from medical devices (e.g.: heart rate, blood pressure respiratory rate) and from subjective sources (e.g.: patient observations).

For example, if the following findings are determined on a patient should the value of the overall score exceed a specified threshold value of 3, as indicated in the first row of Table 4.3. The example illustrated Table 4.4 indicates a total score of 5, which exceeds the limit value of 3.

This is but one example where data from multiple sources are used to create an actionable assessment on a patient. Others exist, particularly in respiratory and cardiovascular system management (e.g.: septicemia models, weaning models).

4.3.3 Mathematical Modeling Techniques Employed in Clinical Decision Making

Many mathematical techniques are employed in clinical decision making. The objective in the following subsections is to provide examples of specific methods to illustrate their utility in application, recognizing that the scope of every method is far beyond the available space afforded this chapter.

Table 4.4 Example rapid response determination on a specific patient based on the ICSI RRT model criteria

Finding	Value	Score
Heart rate (beats/minute)	65	0
Systolic blood pressure (mmHg)	80	2
Respiratory rate (breaths/minute)	8	1
4-h urine output (milliliters)	200	0
Central nervous system	Confusion	1
Oxygen saturation (%)	90%	0
Respiratory support/oxygen therapy	O_2 therapy	1
Total	–	5

Mathematical modeling techniques can include the following:

- Protocols and Algorithms: processes or rule sets that can support automated execution by a computer.
- Signal analysis: signal frequency determination, changes in signal frequency, and power spectral density are often used to determine changes in heart rate, respiratory rate or other periodic signal conditions.
- Simple thresholds: fixed or set numeric value boundaries that define a point at which actions in the form of notifications or alarm signal communications are initiated based upon detecting when specific measurements, or a single measurement, breach a specified boundary or boundaries.
- Ratio calculations: simple ratios of measured or set parameters that provide a measure of performance and to which a simple fixed threshold or trend can be applied. Examples include the rapid-shallow-breathing index (RSBI), which is the ratio of the patient respiratory rate (RR) to the tidal volume (Vt), or modified shock index (MSI), which is the ratio of heart rate (HR) to mean blood pressure (MBP).
- Expert systems: algorithms that facilitate decision making via logical and empirically-defined conditions. These conditions may incorporate artificial intelligence (AI) or may be simple logical statements. For example, if-then-else conditions that consider specific events in succession that, when satisfied, trigger alarm signal conditions.
- Artificial neural networks (ANNs): computational models involving interconnected nodes trained on empirical data to respond with a specific output result to specific input parameter values. ANNs can be trained on a population set of data and then used with a specific patient. The degree to which the ANN produces a result on an individual patient consistent with the behavior of the population is dependent upon how closely the individual patient's specific set of measurements fall within the set limits of the population data. The main point is that through training, the ANN can "learn" and its performance can improve as more data are

employed in its training. A key issue with the use of ANNs is the ability of the ANN to accurately represent the specific patient can be in error, particularly if the patient is not properly represented by the population data. Hence, in this sense, the ANN can be misused.

- Regression Techniques: methods including linear least squares, logistic, polynomial, and stepwise regression are used in predictive modeling between dependent and independent variables in time series. Regression can be applied to time-based data or to investigate the relationship between two or more parameters. The basic objective is to describe in a least-squares sense, the relationship between an input variable, x, and an output variable, y, such that the error or residual between any measurement and the "best-fit" model is minimized.

These various techniques can and have been employed within clinical software, such as clinical decision support systems built into EHRs. Furthermore, these techniques can and have been employed in medical devices for the purpose of diagnosis or identification of specific conditions based upon the collected data.

4.4 Signal Processing and Alarm Signal Annunciation Methods

Examples of applications of various mathematical techniques and methods are provided in this section, identifying the potential uses and the key measures or benefits of each type of mathematical method for the specific application.

4.4.1 Signal Frequency Determination: Fourier Transforms and Lomb-Scargle Periodograms

Signal processing analysis is quite commonly used in medical device data capture. Signal frequency determination is used with electrocardiogram (ECG) data, respiratory rate data, and any automated data capture scenario where analysis of the underlying behavior needs to be understood. It is necessary to determine the frequency or frequency response from analog signal measurements which are transformed into digital signals through sampling. The traditional approach for accomplishing this is the Discrete Time Fourier Transform (DTFT) of the time-sampled signals. An efficient method for computing the DTFT is the Fast Fourier Transform (FFT).

The DTFT requires data having evenly sampled intervals in quantities that follow the 2^N power law. Furthermore, the DTFT requires even distributions of data. That is, data must sampled in even time intervals and there must be no missing data points. For this reason, another method, the Lomb-Scargle Periodogram (LSP), which exhibits invariance with time translation, can effectively deal with empirical issues such as missing data points and unevenly-collected data points [18–20].

The following subsections provide examples of the application of both the DTFT and the LSP approaches to machine-based data, with specific real-world applications.

4.4.1.1 Discrete-Time Fourier Transform and the Fast Fourier Transform

The Discrete Time Fourier Transform (DTFT) based on a time sample of data, $x(t)$, where the discrete time signal, $x[0], x[1], x[2], \ldots, x[N-1]$, is distributed evenly over N samples. The DTFT for any frequency ω is given by Eq. 4.1:

$$X\left(e^{j\omega}\right) = \sum_{n=0}^{N-1} x[n]e^{-j\omega n} \tag{4.1}$$

The individual samples, n, are evaluated for each frequency, ω. Equation 4.1 can be enumerated rather efficiently by way of the Fast Fourier Transform (FFT). The FFT is developed from the DTFT in the following manner. First, the right-hand side of Eq. 4.1 can be re-written for discrete frequencies, where

$$e^{-j\omega n} = e^{-2\pi jnk/N} \tag{4.2}$$

In Eq. 4.2, the equivalence may be made in the exponents so that:

$$\omega = 2\pi nk/N \tag{4.3}$$

From Eqs. 4.1 and 4.3, the Cooley-Tukey FFT algorithm may be developed. This decomposes the original DTFT into two separate transforms of length $N/2$, which are then linearly superimposed as follows:

$$X\left(e^{j\omega}\right) = \sum_{n=0}^{N-1} x[n]e^{-\frac{2\pi jnk}{N}} = \sum_{n=0}^{\left(\frac{N}{2}\right)-1} x[2n]e^{-\frac{2\pi j(2n)k}{N}} + \sum_{n=0}^{\left(\frac{N}{2}\right)-1} x[2n+1]e^{-\frac{2\pi j(2n+1)k}{N}} \tag{4.4}$$

The FFT in Eq. 4.4 can be enumerated in a number of ways, including through Microsoft® Excel® spreadsheet software by way of the *Analysis ToolPak* add-in, as well as through other tools, including Matlab by Mathworks®, IBM SPSS Software, and others.

As an example, consider the time signal given in Eq. 4.5 and illustrated in Fig. 4.6:

$$x(t) = \sin(2\pi f_1 t) + \sin(2\pi f_2 t) + \sin(2\pi f_3 t) + \sin(2\pi f_4 t) \tag{4.5}$$

In Eq. 4.5, the following frequencies are set in the sample time series:

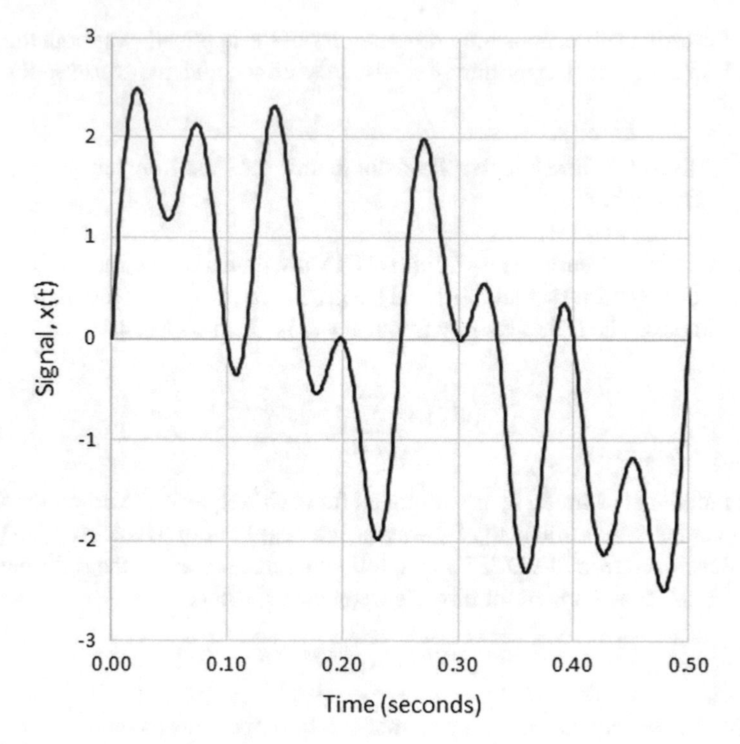

Fig. 4.6 Example time-varying signal with time for FFT analysis

$$f_1 = 2\,\text{Hz} \tag{4.6}$$

$$f_2 = 4\,\text{Hz} \tag{4.7}$$

$$f_3 = 8\,\text{Hz} \tag{4.8}$$

$$f_4 = 16\,\text{Hz} \tag{4.9}$$

The FFT of this time signal extracts the fundamental frequencies, as illustrated in Fig. 4.7.

The peaks in the signal power representative of the output of the FFT computation illustrate the four fundamental frequencies identified in the original time signal, Eqs. 4.6–4.9. The discussion will now turn to reconstructing this result using the LSP.

4.4.1.2 Lomb-Scargle Periodogram (LSP)

The use of the Lomb-Scargle Periodogram (LSP) for the analysis of biological signal rhythms has been well-documented [21, 22]. A benefit of LSP over DTFT is that LSP can effectively be used where missing data points or variable time sampling exists.

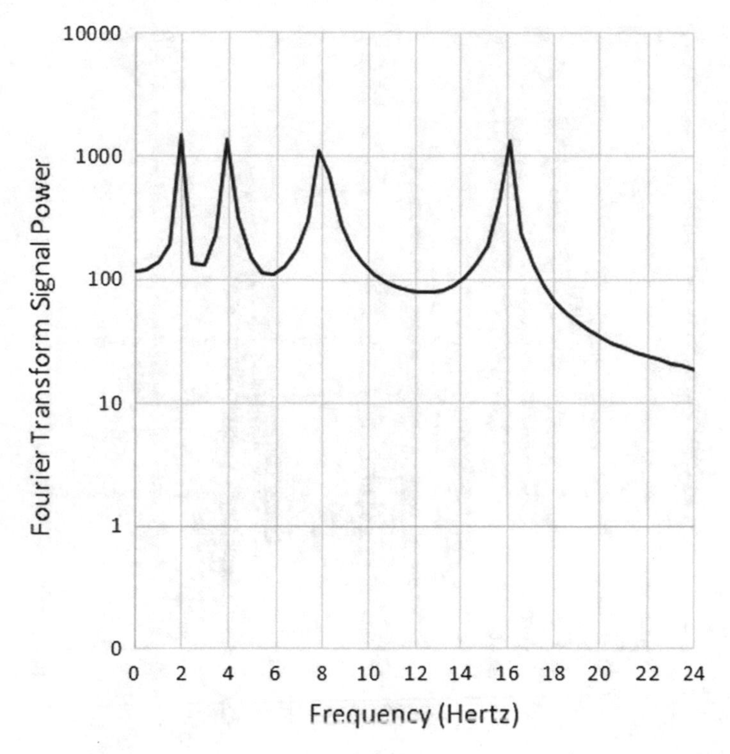

Fig. 4.7 Fourier Transform of time signal presented in Eq. 4.5, illustrating four base frequencies

Thus, the LSP method for periodic signal determination can often be used where the DTFT has difficulty or may not be suitable [21, 22]. The LSP estimates frequency spectrum via a least-squares fit to sinusoids [23]. Single frequency estimation, particularly in the case of unevenly sample data, is a significant reason for considering the LSP [24].

The normalized LSP spectral density calculation is determined using Eq. 4.10:

$$P_N(\omega) = \frac{1}{2\sigma^2}\left\{\frac{\left[\sum_j \left(Y_j - \bar{Y}\right)\cos\omega(t_j - \tau)\right]^2}{\sum_j \cos^2\omega(t_j - \tau)} + \frac{\left[\sum_j \left(Y_j - \bar{Y}\right)\sin\omega(t_j - \tau)\right]^2}{\sum_j \sin^2\omega(t_j - \tau)}\right\} \tag{4.10}$$

and

$$\tau = \left(\frac{1}{2\omega}\right)\tan^{-1}\left[\frac{\sum_j \sin 2\omega t_j}{\sum_j \cos 2\omega t_j}\right] \tag{4.11}$$

where $P_N(\omega)$ is the normalized power as a function of angular frequency. The relationship between angular (or circular) frequency and frequency in Hz is given by:

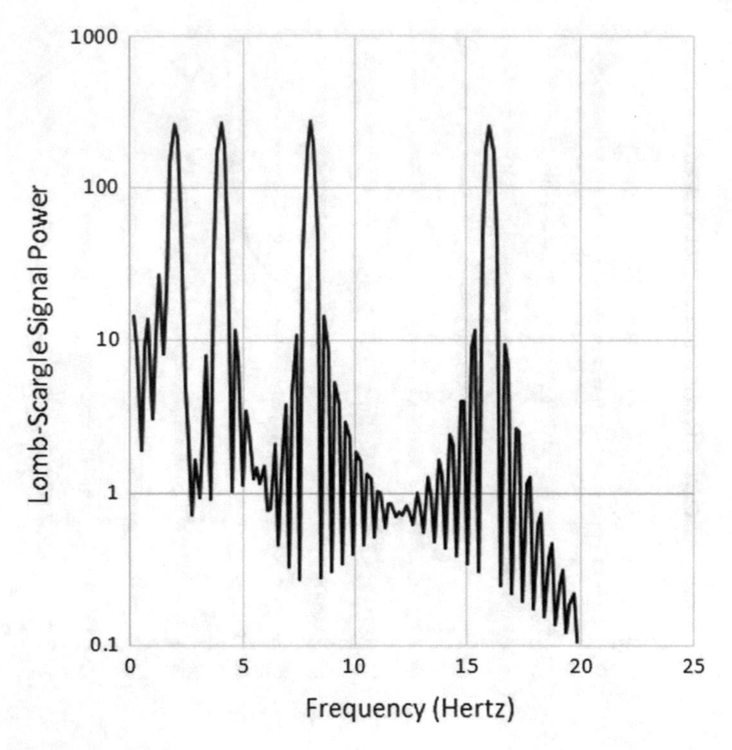

Fig. 4.8 LSP plot of previous sample sinusoid. Method developed as macro in Microsoft Excel

$$f = \frac{\omega}{2\pi} \tag{4.12}$$

The solution of Eqs. 4.10–4.12 is rather straightforward. A method developed in Excel was used to enumerate the example of the previous sub-section, with plot result shown in Fig. 4.8.

This example illustrates that the LSP can produce the same power spectral density as the FFT, with the added advantage that the discrete data are not required to be evenly sampled in time. These techniques can be applied to many applications in medicine pertaining to finding or monitoring frequency domain events, such as heart rate determination or heart rate variability assessment [25], as will be illustrated in the Sect. 4.4.2. The LSP is deemed a better method for evaluating power spectral density in time-varying signals where there may be missing or data gaps, or irregular measurements, which is frequently the case in biology and medicine [21].

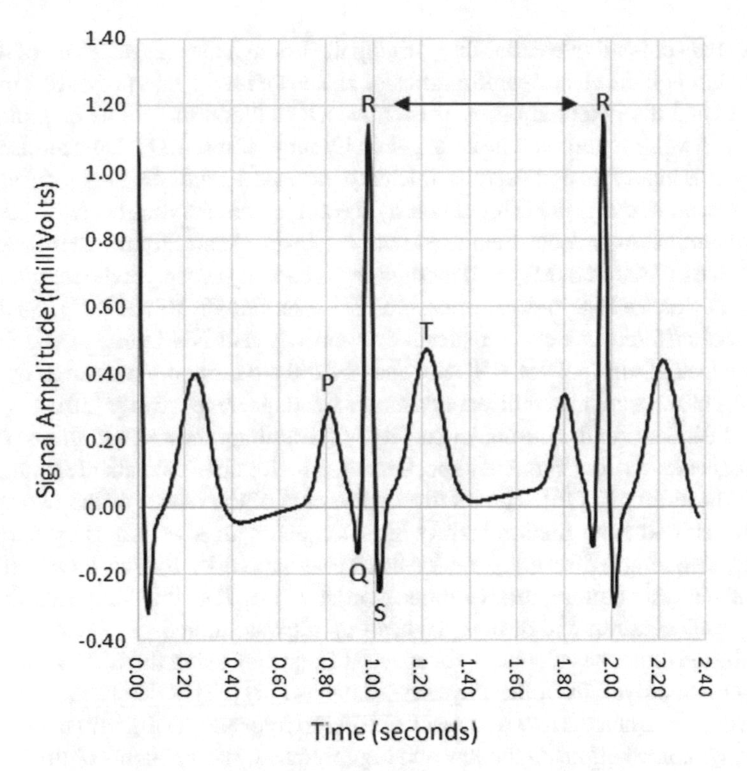

Fig. 4.9 Example electrocardiogram from the multiparameter intelligent monitoring in intensive care II, or MIMIC II, database [28]. Typical Lead II ECG waveform showing the P wave, QRS complex, and T Wave

4.4.2 Application of LSP to Heart Rate Variability Assessment

Heart Rate Variability (HRV) is a measure of the beat-to-beat variability in heart rate, which, in turn, is the reciprocal of the measured R-R intervals of a standard electrocardiogram (ECG). An example of a typical ECG showing the R-R interval is shown in Fig. 4.9.

HRV has been determined to be an adverse event indicator, and a measure of impending decompensation is the standard deviation of the NN (or normal-to-normal) R-R intervals. The standard deviation of the normal R-R interval (SDNN) captured over an ensemble of, say, 24 h has a significance, as well as shorter recordings of 5 min in length [26]. A challenge with the SDNN determination is that measurements need to be made under physiologically stable conditions, prompting shorter time measurements (5 min) in which processing of these shorter intervals is performed after each batch of measurements. Studies have shown that an SDNN of less than 50 ms "is considered indicative of high risk; a SDNN of between 50 and 200 indicates moderate risk; while a value over 100 ms is considered normal" [27].

The study of heart rate variability, principally through the measurement of the R-R interval through the electrocardiogram signal, shown in Fig. 4.9, provides a measure of health and the potential onset of events such as blood infections or septicemia. "HRV of a well-conditioned heart is generally large at rest…During exercise, HRV decreases as heart rate and exercise intensity increase…[and] decreases during periods of mental stress….[and] regulated by the autonomic nervous system" [29].

Furthermore, in patients diagnosed with congestive heart failure (CHF) and acute myocardial infarction (AMI), two methods are generally accepted: the standard deviation of all normal R-R intervals measured between sinus beats (SDNN) and the percentage of differences between successive and adjacent NN intervals which differ by more than 50 ms (pNN50). SDNN and pNN50 are normally measured over 24 h and units of measure are milliseconds and percentage, respectively [30].

The Task Force of the European Society of Cardiology showed variations between high frequency and low frequency spectral densities for different patient attitudes (rest and 90° head-up tilt) [26]. The relative magnitude of the power of the two spectral densities shifted from predominantly high frequency to low frequency during the process, signaling a discriminator for heart rate variability for these two attitudes. Spectral analysis (autoregressive model, order 12) of R-R interval variability in a healthy subject at rest and during 90° head-up tilt was studied.

To demonstrate the calculation of spectral frequency using the LSP, some sample data were employed from the Physionet database [31]. This database contains the ECG waveforms and the low frequency (LF)/high frequency (HF) ratio (a measure of the relative contributions of the low and high frequency components of the R-R time series to total heart rate variability). A plot of a subset of the R-R intervals together with the mean and standard deviations are shown in Fig. 4.10.

The signal is synthesized using a user-definable mean heart rate, number of beats, and sampling frequency, as well as a definable waveform morphology. The plot in Fig. 4.10 shows the beat-to-beat variability, with the mean and the mean $\pm$ the sample standard deviation overlaid.

The corresponding LSP illustrates the principal frequency of the ECG (R-R interval), associated with the highest power, which is 1 s, as shown in Fig. 4.11. This lowest frequency component is 60 beats per minute and is consistent with the actual heart rate.

The LSP is an effective tool for classifying frequencies of unevenly-sampled time-varying signals [32, 33]. These characteristics apply to data collected from medical devices at the point of care, where gaps in data can occur frequently. The use of the LSP for discriminating signal frequencies is be effective and bears further investigation with periodic signal analysis, such as HRV, respiratory rate variability (RRV), wherein frequency and temporal changes are correlated to decompensating condition.

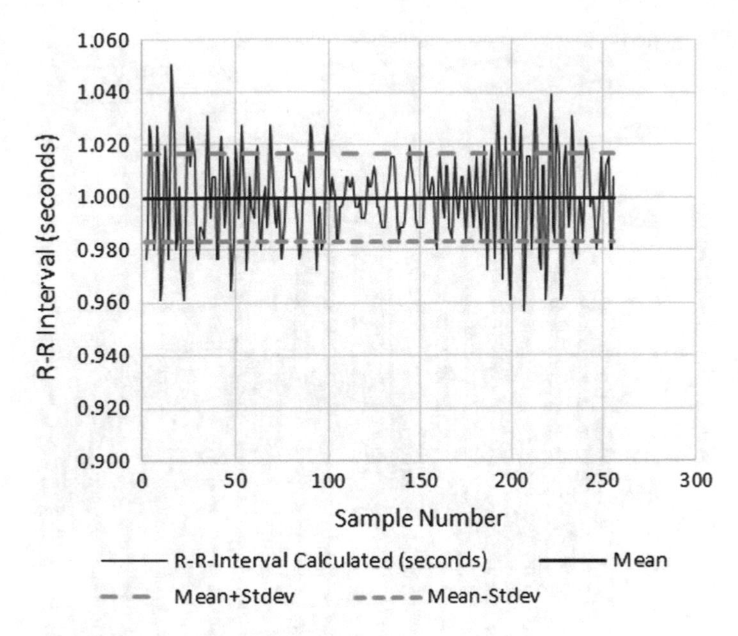

Fig. 4.10 Heart rate signal variability plotted from Physionet data

4.4.3 Combining Data from Multivariate Sources to Arrive at Complex Models of Patient Behavior

There are numerous examples of data combinations from multiple sources that can be derived from both continuous and discrete data. Examples include:

- Sepsis: many models of early onset of sepsis include combinations of HRV, laboratory measurements of blood chemistry such as sodium and potassium levels, temperature increases, R-R variability, and modified shock index (MSI) [34–38].
- Weaning from post-operative mechanical ventilation: rapid-shallow breathing index; high peak pressure combined with low oxygen saturation; low end-tidal carbon dioxide measurement combined with low spontaneous RR [39].
- Obstructive sleep apnea: respiratory depression and failure [40, 41].

Many other examples exist. These examples involve the combining of multiple source data to create indices, similar to the RRT calculation example of Sect. 4.1.2.2. As in the example of that section, the identification of an issue may not necessarily follow a binary pattern (i.e.: disease present or not), but may follow a fuzzy pattern of severity level in which more or less concern (e.g.: low, medium, high priority) is determined based on the combining of multiple parameters to arrive at a representation of severity. In this section certain mathematical techniques are investigated to combine, filter, and define thresholds in time series data for clinical decision making.

Fuzzy logic can be applied to evaluate empirical estimates of severity of illness of condition. One of the benefits achieved in using fuzzy logic is in the reduction of

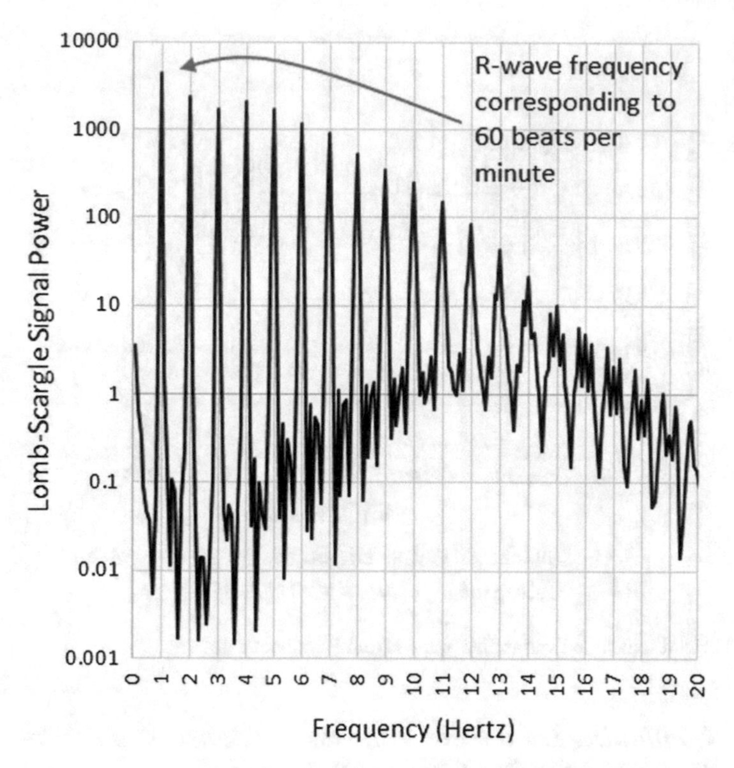

Fig. 4.11 LSP calculation showing the principal frequency with heart rate of 60 beats per minute along with harmonics frequency components

false alarms. The challenge of reducing false alarms is one of achieving a balance between identifying the essential, actionable information with which to notify clinical stakeholders so that they may intervene on the behalf of the patient to prevent or head off adverse events that can lead to death, from the spurious, non-emergent and non-actionable events that are not indicative of patient harm. The issue is that in the process of reducing false alarms to minimize spurious noise that can create fatigue in clinical stakeholders, thereby making them less effective to identify actual events, the possibility of missing truly emergent events can occur, with concomitant negative or even disastrous effects on the patient.

4.4.4 Decision Making Based on Fixed Thresholds and Fuzzy Logic

Fuzzy logic seeks to apply specific rules to define actions when severity of condition must be taken into account in which the action follows an empirical set of outcomes rather than binary, defined by clear fixed thresholds. Decision methods

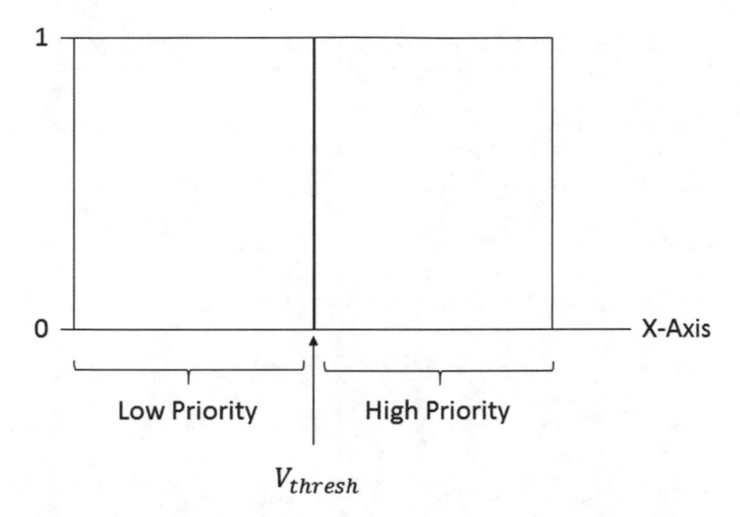

Fig. 4.12 Simple priority threshold: when a value exceeds a fixed threshold, a decision between two binary conditions occurs. In the case here, low priority and high priority are the two binary events

using fixed thresholds can follow simple if-then-else constructions whereby when a value exceeds some specific threshold, an action is performed. A generic and simple example of this is illustrated in Fig. 4.12, in which the transition from a Low Priority to a High Priority condition (and back again) is determined based upon a measured value on the X-Axis exceeding (or dropping below) a fixed threshold value given by the variable V_{thresh}. This type of analysis and decision making can serve adequately in many applications. Examples from everyday life include the light sensors in an automobile that turn the headlights on and off based upon the relative ambient light level (i.e.: lack of light) outside a vehicle. Another example is the refrigerator light which turns on when the door is opened and off when door is closed (or so we are led to believe).

Yet, in life there are usually shades of gray: situations that are not clearly delineated by a simple threshold. In patient monitoring, thresholds (both fixed and fuzzy) come into play in the annunciation of alarms associated with measurements falling outside of normal, safe ranges. Examples of alarms that can impact patient safety if they are allowed to persist include Asystole (i.e.: lack of heart beat), low blood oxygenation level (i.e.: measurement of low oxygen levels in blood hemoglobin), high peak pressure (i.e.: high pressure measured in the tubing of patients being mechanically ventilated), low or high end tidal carbon dioxide measurement (i.e.: hypo- and hypercarbia, indicating problems in respiratory drive and depression, possibly related to obstructive or central sleep apnea), and the list goes on.

Yet, in situations as listed above, it is frequently the case that the breach of a simple threshold is not an indicator of a problem requiring intervention, but rather an indicator of a false alarm. To illustrate this, consider the diagram of Fig. 4.13. A fixed threshold defines an operating point at which a certain number of true, actionable

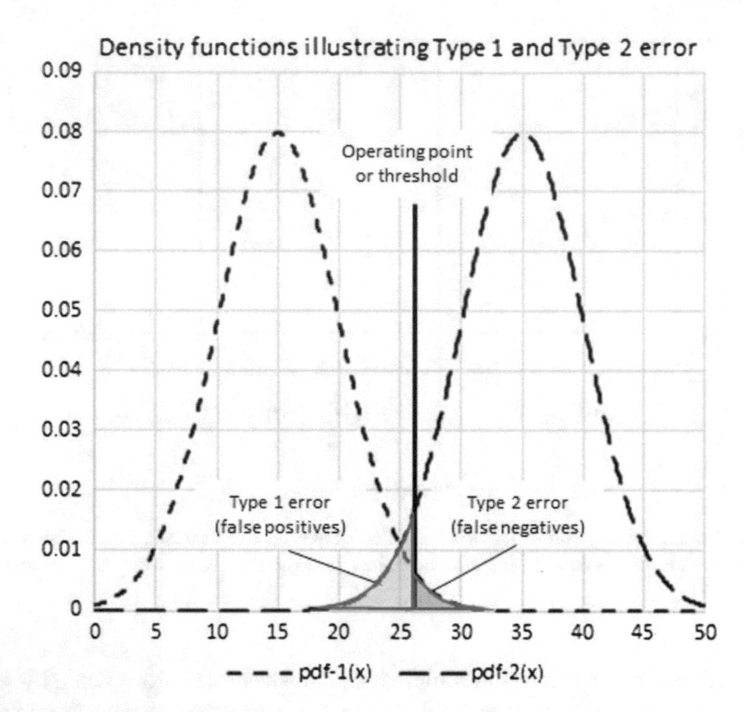

Fig. 4.13 Type 1 and Type 2 error illustration example showing probability density functions 1 and 2 (pdf1, pdf2) with a set operating point or threshold

events and false, non-actionable events may occur. Discriminating between true positive events and false positive events is the challenge, wherein the cost of reducing false positive events may be the occurrence of false negative events.

In medicine, the phenomenon of alarm signal fatigue is a key source of patient safety risk: non-actionable alarms caused by patient movement, sensor anomalies and other artifacts can cause the issuing of alarms based on false positive events. While not dangerous to the patient, these false positive events can result in the clinician failing to recognize true positive events when they occur, which can impact the ability of a clinician to respond to the patient when a real alarm signal is issued. Clinicians can become insensitive to the alarms as most are non-actionable. Estimates place the number of non-actionable, or clinically-insignificant alarms that inundate care providers "with false positive rates of 85–99%" [2, p. 13].

By regulating when alarm signals are issued through the incorporation of empirical data, such as delays associated with more than some specified quantity of persistent or non-self-correcting measurements, the overall quantity of issued alarm signals can be reduced. The understanding of false alarms can be expressed mathematically through the study of Type 1 and Type 2 error. The difference between Type 1 and Type 2 errors are illustrated with the aid of Table 4.5.

The concept of Type 1 and Type 2 error can be illustrated qualitatively with the aid of Fig. 4.13. The operating point or threshold can correspond to a fixed level at

Table 4.5 Illustration of type 1 and type 2 error

		Actual or real	
		True	False
Measured or perceived	True	Correct	Type 1 -false positive
	False	Type 2—false negative	Correct

which action is to be taken once breached. The figure depicts two probability density functions (pdfs), one of which (pdf1) corresponds to true events, and the other (pdf2) corresponds to false events. That is, pdf1 is a distribution of measurements which are known to occur when true events are detected, whereas pdf2 is a distribution of measurements which are known to occur when no adverse events are detected. The overlap between the two distributions is the focal point of interest.

As is illustrated, in order to minimize the Type 1 error, there may be a trade in the quantity of Type 2 errors, or false negatives that must be accepted. This is particularly the case when the two distributions overlap one another, as shown here. In order to reduce the quantity of true negatives—desirable for the patient, as this will reduce the likelihood of missing adverse events—one must necessarily accept more false positive events. Conversely, to reduce false positive events—desirable for the clinician, as this will reduce the quantity of non-actionable events—one must necessarily accept more false negative events. Selection of the optimal threshold is the challenge. Mathematically, selection of this threshold to minimize the null hypothesis when it is true (i.e.: Type 1 error) and failing to reject the null hypothesis when the alternate hypothesis is true (i.e.: Type 2 error) are defined as the thresholds α and β, respectively. The value of α often selected as 0.05 [42].

The dilemma: how to maximize the likelihood of detecting true positive events while minimizing non-actionable events? A fixed threshold in such a situation may not be very effective as the quantities of false alarm signals can often number into the thousands [2, p. 13].

Therefore, the concept of introducing fuzzy logic is motivated, and with it the concept of combinatorial methods that incorporate multiple data sources to provide a more complete and contextually accurate view of the patient and condition at hand.

An example of a fuzzy set is represented in the diagram of Fig. 4.14. In this example membership in three different membership functions (MF)—Low, Medium and High priority—is defined by the relative value along the X-Axis (arbitrary, for this example). The relative value and, thus, the association level, is defined by the X-Axis position in relation to the triangle membership functions. The overall value of fuzzy set membership is defined by the test value selected along the X-Axis.

The Low Priority membership function is defined as follows:

$$\mu_L(x) = \begin{cases} 1, x < V_L \\ 1 - \frac{x-V_L}{V_M-V_L}, V_L \le x \le V_M \\ 0, x > V_H \end{cases} \tag{4.13}$$

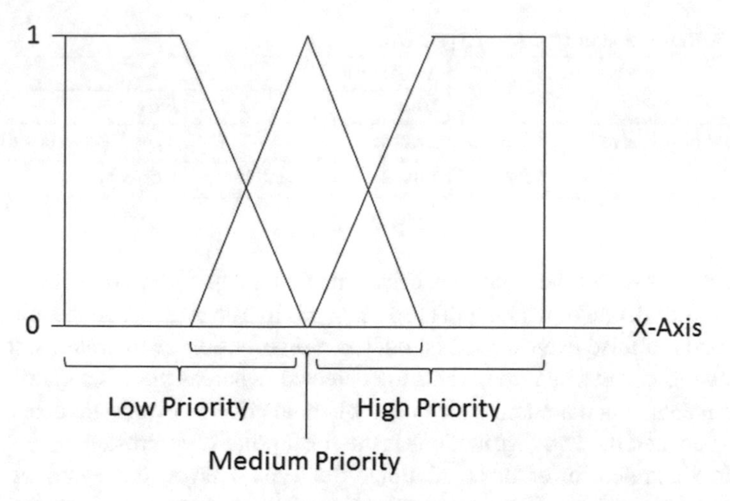

Fig. 4.14 Fuzzy sets showing a low priority, medium priority and high priority set of membership functions (MF). Membership to any set is defined by the relative position along the X-axis, identifying the relative association with low, medium and high priority states

The Medium Priority membership function is defined as follows:

$$\mu_M(x) = \begin{cases} 0, x < V_L \\ \frac{x - V_L}{V_M - V_L}, V_L \le x \le V_M \\ 1 - \frac{x - V_M}{V_H - V_M}, V_M \le x \le V_H \\ 0, x > V_H \end{cases} \tag{4.14}$$

The High Priority membership function is defined as follows:

$$\mu_H(x) = \begin{cases} 0, x < V_M \\ \frac{x - V_M}{V_H - V_M}, V_M \le x \le V_H \\ 1, x > V_H \end{cases} \tag{4.15}$$

Consider a worked example, illustrated by the sample membership functions shown in Fig. 4.15, with end tidal carbon dioxide ($etCO_2$) as the X-Axis values. Measurement of $etCO_2$ is of particular importance when monitoring patients post-operatively who are diagnosed or at risk for obstructive sleep apnea (OSA) or central sleep apnea (CSA). A key measure of respiratory depression is $etCO_2$, which is measured using a nasal cannula using special monitors. Normal $etCO_2$ for adults ranges from 35 to 45 mm of mercury (mmHg). Certain changes in both $etCO_2$ and RR are indicators of serious illness. For example:

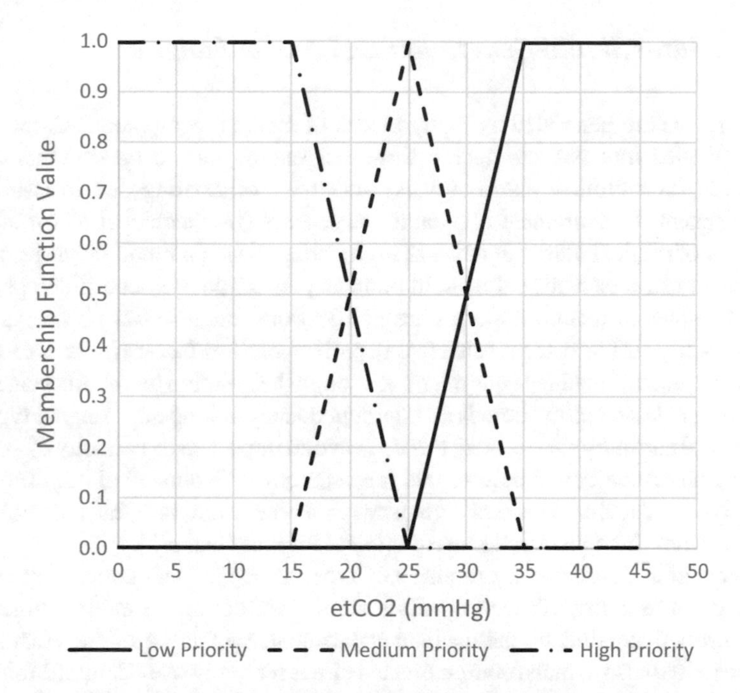

Fig. 4.15 Example membership functions for end tidal carbon dioxide. The X-Axis is end tidal carbon dioxide ($etCO_2$) measured in millimeters of mercury (mmHg)

- Hyperventilation, in which breathing exceeds 24 breaths/minute, is consistent with higher mortality rates and adverse events in hospital critical care and general wards [43]. Similarly, hyperventilation is related to increased carbon dioxide expiration.
- Hypercapnia, in which $etCO_2$ exceeds 45 mmHg, is consistent with respiratory failure. Hypocapnia, in which $etCO_2$ falls below 35 mmHg, is consistent with hypoperfusion, hyperventilation or hypothermia [44].

Physiologic monitors, mechanical ventilators and other medical devices used for diagnosis, therapy and treatment in ORs, ICUs (Intensive Care Units) and lower acuity settings employ the use of thresholds placed on parameters to cause the issuing of alarm signals when values breach actionable. The measure $etCO_2$ must remain within certain limits, otherwise patients may be determined to be at risk. The membership functions shown in Fig. 4.15 illustrate the levels associated with low, medium and high priority levels of $etCO_2$ measurement. An $etCO_2$ below 15 mmHg is deemed a high priority actionable event. An $etCO_2$ value between 15 and 25 mmHg may be a warning level, possibly requiring action. An $etCO_2$ value between 25 and 35 mmHg is deemed a low priority level but will require monitoring should the patient persist within this region continuously.

The benefits of using fuzzy logic versus fixed thresholds is to accommodate the "soft" transitions between states. Thus, the transition from low priority, such as a low priority alarm, versus a high priority alarm.

4.4.5 Alarm Signal Generation and Alarm Fatigue

Alarm signals are generated by many pieces of medical equipment. Alarm signals can be divided into two categories: those originating due to measurements from the patient (i.e.: clinical alarm signals) and those originating due to issues with the equipment, measurement apparatus, or sensors (i.e.: technical alarm signals). Examples of clinical alarm signals are notifications when a blood pressure or heart rate measurement exceeds a threshold boundary for some pre-determined period of time. Examples of technical alarms are notifications that a sensor probe, such as a pulse oximetry cuff, is disconnected from a patient, or a low battery indicator has been issued on a piece of medical equipment. Although these define two distinct categories of alarm signals, in reality there is an interdependency among alarm signal types, for if a technical alarm signal is issued, this can have an impact on the veracity of a clinical alarm signal or can completely negate the accuracy of a clinical alarm. Therefore, certain types of technical alarm signals, such as the premature disconnecting of a sensor, can also be considered to have clinical implications [45].

Clinical alarm signals vary in terms of type, frequency and content by medical device, and the quantity of alarm signals generated and communicated to clinical staff can also vary depending on the medical department. By far, the settings that experience the most alarm signals, communicated either remotely over Central Monitoring Stations or remotely through Nurse Call Systems, Voice-over-IP (VOIP) phones, and smart appliances, is largest in the intensive care units (ICUs), where technologically-dependent patients are most often found:

> Hospital staff are exposed to an average of 350 alarms per bed per day, based on a sample from an intensive care unit at the Johns Hopkins Hospital in Baltimore. [46]

As discussed in Sect. 4.4.4, not all alarm signal carry significant meaning. The quantity of alarm signals issued by multi-parameter monitors, mechanical ventilators and ancillary medical equipment translate into alarm signals to which clinical staff must react. Missing a key clinical alarm signal can be a patient safety hazard. On the other hand, responding to alarm signals that do not carry clinical import, or non-actionable alarm signals, translates into clinician fatigue, which can carry its own source of patient safety risk [2, p. 9].

Differentiating between false, non-actionable alarm signals and true, actionable alarm signals requires bringing both patient and clinician context to bear on the situation as well as innovative mathematical techniques to assist. Among these techniques includes:

- Searching for patterns among alarm data to detect true versus false events;
- Interrogating sustained alarm signal conditions that are non-self-correcting over some pre-defined period of time (i.e.: sustained alarm signals);
- Combining multiple measurements from different sources together to identify correlations in measured behavior that can carry clinical meaning (i.e.: combinatorial alarm signals); and,

- Searching for events that repeat at some pre-defined frequency of occurrence or in combination with other events that occur at some frequency of occurrence (i.e.: combinatorial alarm signals combined with repetitive or trended behavior).

Other approaches involve bringing heuristics to bear that combine empirical evidence associated with specific demographic or patient population statistics together. This approach, however, may be very specific to a class of patient and set of clinical conditions.

In order to illustrate examples of these types of mathematical techniques, application to a specific clinical situation or challenge may inspire the reader to a better understanding. An area of increasing focus is that of post-operative respiratory depression due to the administration of opioid pain medication. This is of particular importance in the case of patients who suffer from obstructive sleep apnea (OSA). An approach for monitoring patients at risk for respiratory depression is capnography, in which end-tidal or exhaled carbon dioxide is measured via nasal cannula. Undetected opioid-induced respiratory depression has also been identified as one of the top 10 patient safety hazards by ECRI [3].

Capnography is seen as a more accurate and rapid measurement of the onset of respiratory depression over the older and more accepted peripheral blood oxygen saturation measurement of pulse oximetry because capnography measures ventilation. Measurements of ventilation provide an instantaneous understanding of respiratory system compromise, whereas pulse oximetry can often provide a delayed reading of a minute or longer [47–50].

In relation to alarm signals and alarm signal fatigue, hospital systems considering the use of capnography and pulse oximetry in the general ward have that almost 9 in 10 hospitals indicated they would increase their use of patient monitoring, particularly of Capnography and pulse oximetry, if false alarms could be reduced. Furthermore, more than 65% of hospitals that have employed pulse oximetry or capnography have experienced reductions in overall adverse events, of which more than 56,000 and 700 patient deaths have been linked to patient-controlled analgesia for which capnography monitoring can play a great role to ameliorate [51].

The problem with attenuating alarm signal data is achieving the balance between communicating the essential, patient-safety specific information that will provide proper notification to clinical staff while minimizing the excess, spurious and non-emergent events that are not indicative of a threat to patient safety. In the absence of contextual information, the option is usually to err on the side of excess because the risk of missing an emergent alarm or notification carries with it the potential for high cost (e.g.: patient harm or death). Capnography provides real-time feedback: when $etCO_2$ is high, this indicates respiratory failure, and metabolic, perfusion, when $etCO_2$ is low [52]. Alarm signal reporting from medical devices is key to alerting clinical staff of events. Yet, the concomitant fatigue of responding to many false alarms may render clinical staff "snow-blind" to real events, or cause the alarms to be ignored or even turned off, obviating any potential benefit.

For illustration purposes, a hypothetical sampling of data based on experiential measurements, will be used as the target for discussion. Figure 4.16 illustrates sim-

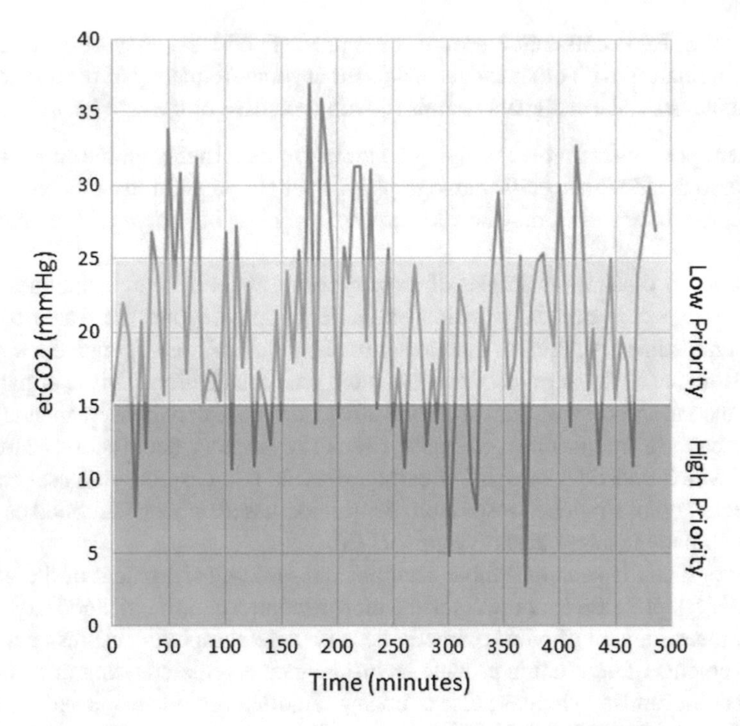

Fig. 4.16 Low and high priority ranges of hypocapnia displayed on top of $etCO_2$ plot

ulated $etCO_2$ versus time for a hypothetical patient. The range and behavior of the data are based on real-data, inclusive of the aberrations and trends of such data. The range of values are also span normal and abnormal as well as emergent ranges. In general, normal values for $etCO_2$ span the range of 35–45 mmHg in adult humans. The low and high priority alarm signal ranges for the data set under consideration in this figure show those values which fall within the low- and high-priority ranges outside of this range as many capnography devices have alarm settings for emergent conditions set to provide notifications below 25 mmHg.

When considering continuous monitoring of $etCO_2$ if alarm signal ranges on the capnography are as identified in the figure, one can see the potential to issue alarms quite frequently. However, many of these alarms are "one-and-done." That is, they occur because of some aberrant behavior (e.g.: shifting of nasal cannula, or patient moving in bed) that cause the measurement to spike or register as an out-of-bounds value with respect to the alarm levels set on the monitor itself.

If these alarm signals are issued at the time they occur, lacking context, one may see that they can provide a frequent source of distraction, particularly for the nursing staff.

This is not to suggest that the chief objection is irritation to the clinical staff. But, from a patient safety perspective, when alarms go off and there is no clear distinction between an emergent alarm signal versus a nuisance alarm signal, this can impact

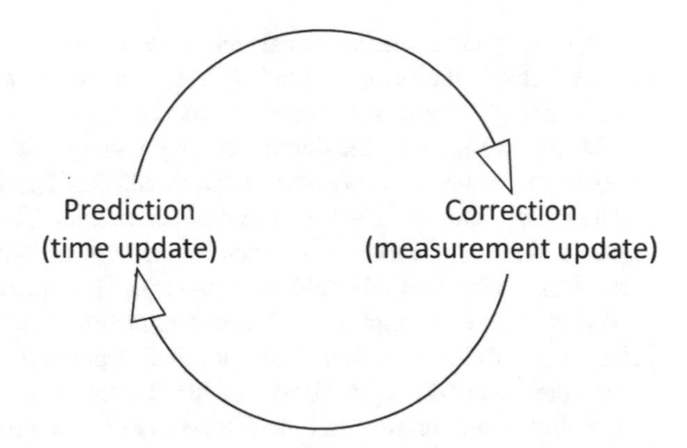

Fig. 4.17 Basic recursive process of the Kalman Filter operation—prediction of future state, or state at future time, based on the best model and knowledge of the current state, followed by a correction based on measurements at that future state

the ability to react appropriately to the patient. Upon inspection it can be seen quite readily that a number of "spikes" wherein measurements are shown dipping down into the "red" area. It can be noticed, too, that in many instances the measurements rise back into a less emergent or normal area upon the very next measurement. Hence, the measurements that would normally trigger the alert do not persist. What is the cause of this? Many possible answers, from artifact in the measurement cannula to movement of the patient.

Problems may be more readily identified if other corroborating information can be brought to bear, such as corresponding changes in respiratory rate, heart rate, SpO2 values. Moreover, if problems are truly present, it would be logical to conclude, based on experience and policy that these "events" would either persist, or would increase in frequency over time. In other words, to verify that behavior are not merely incidental but are correlated to some behavior, the expectation of continued depression or trending depression towards hypocapnia or hypercapnia would be present. A simple way of measuring this (given no other information) is a sequence of measurements at or around the emergent value. For instance, multiple measurements over a period of time of, say, 20 or 30 or 40 s in which the values are depressed or elevated. Or, a series of spikes that occur rapidly over a fixed period of time. In reviewing the data retrospectively, and in light of the desire to reduce spurious notifications, several methods will be considered for determining the viability of reducing or filtering out such measurements.

One method that is particularly well-defined for smoothing and filtering of noisy signals is Kalman Filtering (KF). Kalman filtering is an optimal methodology in the least-squares sense, by estimating the state of a discrete-time process and then correcting the estimate based on future measurements. In this way, the KF is a form of feedback control [53]. The process can be represented with the aid of Fig. 4.17.

The Extended Kalman Filer (EKF) is also a discrete-time process. But, unlike the DTKF in which the process is defined by a linear stochastic difference equation, the EKF seeks to linearize about the current mean value and covariance. When processes being estimated are nonlinear, then the EKF demonstrates its utility [53].

Expanding on the mathematical techniques employed, another reason for filtering of data includes the smoothing of artifact or spikes that are due to signal errors or other issues associated with signal acquisition.

The following examples demonstrates the use of the Kalman Filter in tracking time-varying signals. The concept of optimal filtering of data has many advocates and many applications. The use of formalized methods, including least-squares techniques and the application of Kalman Filtering to the study of medical signals and other data has been widely published [54–56], [12, pp. 169–176].

Kalman filtering employs a recursive algorithm that is an optimal estimator to infer and establish an estimate of a particular parameter based upon uncertain or inaccurate observations. A benefit of the Kalman filter is that it allows recursive processing of measurements in real-time as observations are made and, so, can be applied to live data readily. The Kalman filter also provides for tuning and filtering to enable removal or attenuation of noise.

The generalized equations defining the Discrete-Time Kalman Filter (DTKF) are the state estimate and the measurement update:

$$x_k = Ax_{k-1} + Bu_k + w_{k-1} \tag{4.16}$$

$$z_k = Hx_k + v_k \tag{4.17}$$

- x_k is the state vector containing the estimate of the future state at time k based on previous state at time $k - 1$ ("k minus 1");
- u_t is the vector containing any control inputs;
- A is the state transition matrix which applies the effect of each system state parameter at time $k - 1$ on the system state at time k;
- B is the control input matrix which applies the effect of each control input parameter in the vector u_k on the state vector;
- w_{k-1} is the vector containing the process noise terms at time $k - 1$ for each parameter in the state vector. The process noise is assumed to be drawn from a zero mean multi-variate normal distribution with covariance given by the covariance matrix Q_k;
- z_k is the vector of measurements at time k;
- H is the transformation matrix that maps the state vector parameters into the measurement domain; and,
- v_k is the vector containing the measurement noise terms for each observation in the measurement vector. The measurement noise is assumed to be zero mean Gaussian noise with covariance R.

The filter solution balances the confidence in the measurements with the confidence in the estimates. That is, the filter will respond more closely to the measurements if the "belief" or confidence in the measurements is greater than the confidence in the estimates, and vice versa. If the measurement noise is high, the confidence in the measurements is fairly low, and the filter will smooth out the transitions between measurements, resulting in a state estimate which is less perturbed but also that does not react to sudden changes in measurements (hence, less likely to react to sudden or spurious changes).

The solution process begins with a time update, or prediction. The prediction equations are:

$$\hat{x}_{k|-} = A\hat{x}_{k-1} + Bu_k \quad (4.18)$$

$$P_{k|-} = AP_{k-1}A^T + Q \quad (4.19)$$

Equation 4.18 is the future state predicted to the next time step. Equation 4.19 is the covariance update projected to the next time step.

The prediction is followed by a correction step, or measurement update, in which measurements are used to update the state and correct the prediction:

$$K_k = P_{k|-}H^T\left(HP_{k|-}H^T + R\right)^{-1} \quad (4.20)$$

$$\hat{x}_k = \hat{x}_{k|-} + K_k\left(z_k - H\hat{x}_{k|-}\right) \quad (4.21)$$

$$P_k = (I - K_k H)P_{k|-} \quad (4.22)$$

The solution proceeds with an initial guess on the covariance, P_k, and the state estimate, $\hat{x}_k$. The minus sign (k|-) indicates the estimate at the previous iteration before update occurs. When the assumptions and particulars of the signal are applied to these specific equations, they reduce to the following.

Time update:

$$\hat{x}_{k|-} = \hat{x}_{k-1} + u_k \quad (4.23)$$

$$P_{k|-} = P_{k-1} + Q \quad (4.24)$$

Measurement update:

$$K_k = P_{k|-}\left(P_{k|-} + R\right)^{-1} \quad (4.25)$$

$$\hat{x}_k = \hat{x}_{k|-} + K_k\left(z_k - \hat{x}_{k|-}\right) \quad (4.26)$$

$$P_k = (I - K_k)P_{k|-} \quad (4.27)$$

Some definitions:

- K_k is the Kalman gain;
- P_k is the state covariance matrix

Table 4.6 shows the process for the first 6 measurements, with the assumptions that:

$$P_0 = 100$$

$$R = 1$$

Table 4.6 Kalman Filter model developed in Microsoft® Excel®

Time (sec)	End-tidal CO_2 measurement: (mmHg) versus time (seconds)	vk (measurement noise)	qk (process noise)	Pk (covariance)	End-tidal CO_2 track (mmHg)	Pk-minus	Xk-minus	Kk (Kalman Gain)
0	15.42	1	0.1	1.000	0.000	1.100	0.000	0.524
5	19.34	1	0.1	0.524	10.129	1.100	0.000	0.524
10	22.00	1	0.1	0.384	14.691	0.624	10.129	0.384
15	19.57	1	0.1	0.326	16.284	0.484	14.691	0.326
20	7.49	1	0.1	0.299	13.655	0.426	16.284	0.299
25	20.72	1	0.1	0.285	15.670	0.399	13.655	0.285
30	12.15	1	0.1	0.278	14.690	0.385	15.670	0.278

$$Q = 0.01$$

$$x_0 = 0.$$

A plot of the state estimate of the signal measurements is provided in Fig. 4.18. Note that the estimate follows a nominally mean path relatively unaffected by the noisy of the measurements. This is not to imply that this result is a measure of "goodness": the lack of responsiveness can translate into incorrectly masking real events. The tracking signal illustrates, depending on the level of process noise allowed in the filter, the relative smoothness of the fit (that is, following the general trend) versus following each measurement. This balance between tracking "random noise" and smoothing out the signal is one of the key benefits offered by an optimal algorithm like the Kalman Filter: spikes and aberrations that might otherwise result in the issuing of spurious alarm signals can be smoothed out of the signal.

4.4.6 *Time Averaging*

Time averaging is a special case of Kalman Filtering, and involves calculation of averages over the running sample. For example, if measurements are collected every 5 s, and the running sample is over 20 s, then (including the first sample), a total of 5 measurements would be sampled together. The running average, with weighting

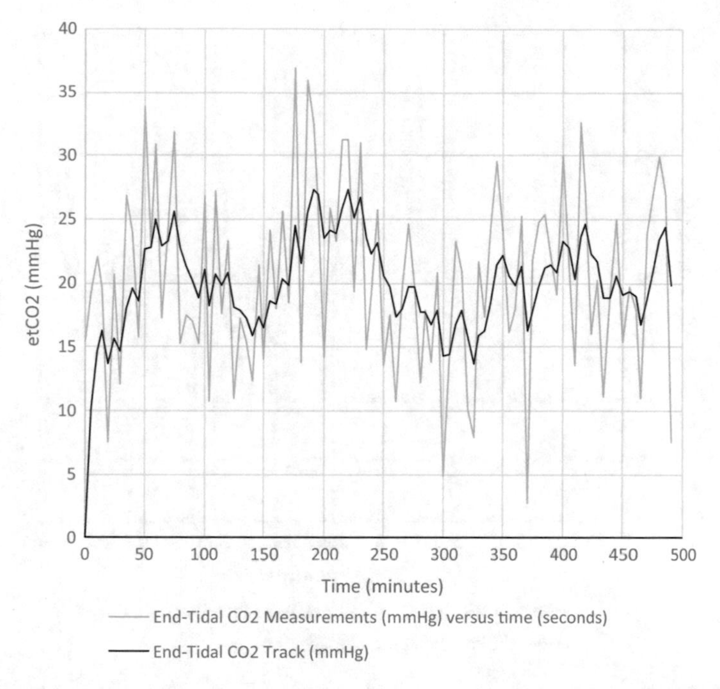

Fig. 4.18 Kalman filter tracking of end-tidal carbon dioxide with measurement error set to 1 mmHg

function $f(i)$ a measure of the relative weight of each measurement, is given by Eq. 4.28:

$$A(t_k) = \sum_{i=k-N}^{i=k} f(i) \times A(t_i) \tag{4.28}$$

where $f(i)$ is the fractional weight associated with each measurement $A(t_i)$.

In the specific case of 20 s signal averaging, N = 5 (t = 0, 5, 10, 15, 20). For equal weighting of the measurements,

$$f(i) = \frac{1}{N}, \quad and \tag{4.29}$$

$$\sum_{i=k-N}^{i=k} f(i) = 1. \tag{4.30}$$

Figures 4.19 and 4.20, respectively, illustrate the running averages associated with 20-s and 30-s running average intervals.

One possible concern with time averaging is that the current value is based on the average of some preceding set, which can mask the current value and delay reporting. The delayed reporting can be both beneficial as well as problematic in that a delay can

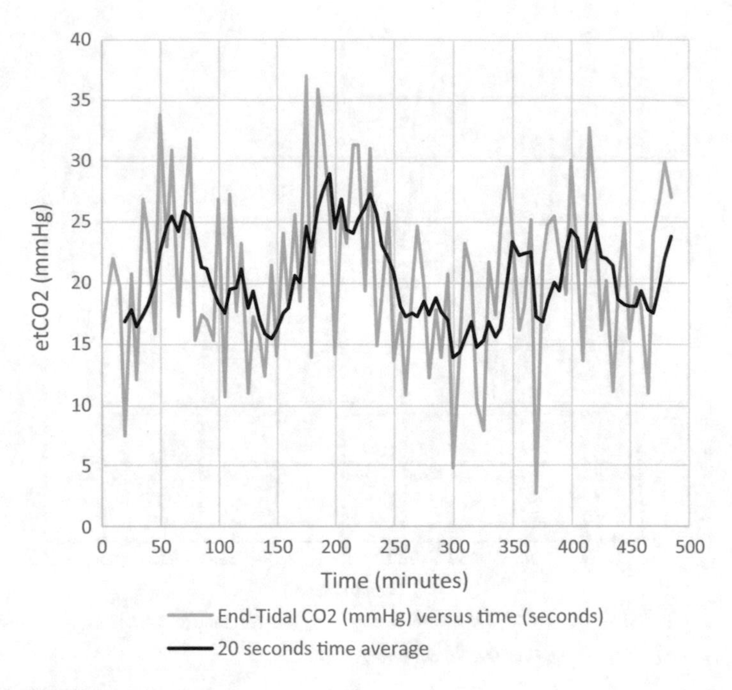

Fig. 4.19 Signal with 20 s time averaging overlay

help to reduce the effect of spurious measurements. Yet, the averaging can also result in delayed reporting of instantaneous events which may be critical (e.g.: no heart beat or no respiratory rate). Balance and selection based on the specific application must be undertaken to determine the best approach.

4.4.7 Transforms: Fourier and Wavelet and Application to Threshold Identification and Data Reconstruction

In the earlier discussion on Fourier series, detection of periodicity of the signal frequency was reviewed and a comparison was made relative to the Lomb-Scargle Periodogram. These series calculations endeavor to transform the data from a time-series basis to another basis (frequency), for the purpose of extracting and increasing visibility into characteristics that are not readily able to be seen in the original basis.

The Fourier transform is a frequently-used and popular signal processing method for transforming bases. Developed by Joseph Fourier in 1807, the essence of the Fourier transform is that any given periodic signal may be represented as the superposition of sine and cosines [57]. Extended to aperiodic signals, the superposition (or sum) is represented as a complex integral of the form:

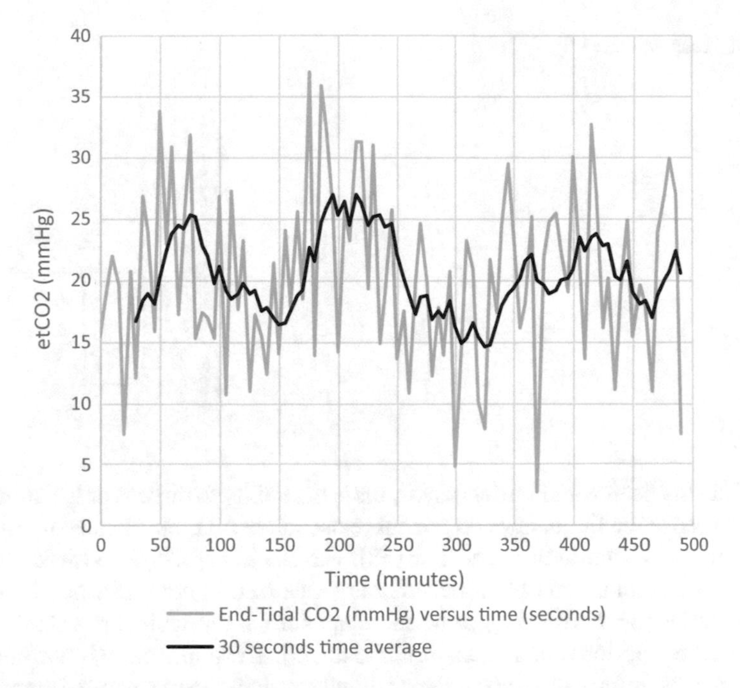

Fig. 4.20 Signal with 30 s time averaging overlay

$$X(f) = \int_{-\infty}^{\infty} x(t)e^{-i2\pi ft}dt \tag{4.31}$$

where $x(t)$ is the time-based signal; f is the frequency; and, $X(f)$ is the transformed signal into the frequency domain.

This continuous form of the Fourier transform can be written as a discrete transform known as the discrete time Fourier transform (DTFT) in which data are sampled over N intervals over a total period of time, T. The time increment per interval, ΔT, then represents the increment between samples. Re-writing:

$$DFT(f_n) = \frac{1}{N}\sum_{k=0}^{N-1} x_k e^{-i2\pi f_n k\Delta T} \tag{4.32}$$

where f_n is the discrete frequency sample, and x_k is the sampled time-series measurement.

The discrete-time inverse transform is then defined as follows:

$$x_k = \frac{1}{\Delta T}\sum_{f_n=0}^{(N-1)/T} DFT(f_n)e^{i2\pi f_n k\Delta T} \tag{4.33}$$

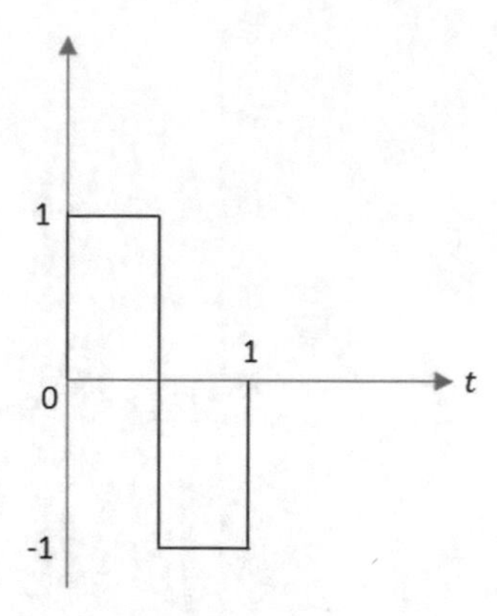

Fig. 4.21 Haar wavelet

A difficulty seen with Fourier transforms is the ability to discern or isolate a signal in time because the frequency components exist across all time. Hence, time and frequency are not visible at the same time [58]. Wavelet transform deconstructs a signal in terms of a summation of wavelets that are localized in both time and frequency. Thus, wavelet transforms tend to be more apt for (in particular) representation of transitional or spurious signal transformations [59]. The implication is that both time and frequency information are preserved and available using wavelet transforms, making them particularly useful for situations where the occurrence of particular events is important information.

The continuous wavelet transform (CWT) maps a sequence of numbers to a new basis. But, unlike the Fourier transform, does not require the signal to be finite in duration or periodic [60, p. 9]. Thus, if $a, b \in R$ are real, continues variables, then the CWT is given by:

$$F(a, b) = \int f(t)w\left(\frac{t-a}{b}\right)dt \tag{4.34}$$

where $w\left(\frac{t-a}{b}\right)$ is the wavelet and $f(t)$ is the time signal. Then, the inverse transform is given by:

$$f(t) = \iint F(a, b)w\left(\frac{t-a}{b}\right)dadb \tag{4.35}$$

The simplest of wavelets is the Haar wavelet, represented in the diagram of Fig. 4.21. The details of the Haar basis and Haar wavelet transform are available elsewhere [60, p. 26, 61].

The wavelet is generated over N data elements, corresponding to the number of data points in the discrete time series measurement sample. In the case of the Haar wavelet transform, the Haar matrix, W_N, with $N = 4$, this becomes:

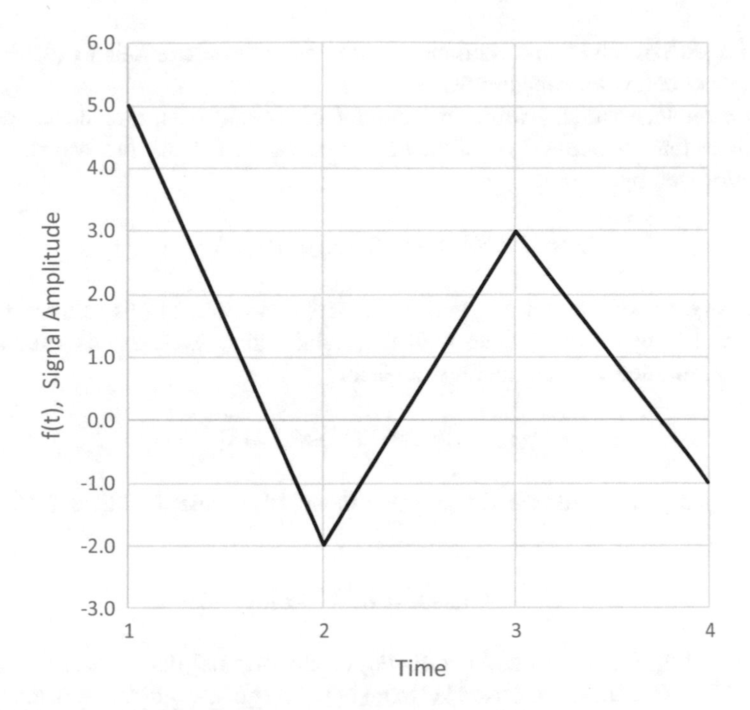

Fig. 4.22 Time signal, $f(t)$, for Haar wavelet transform example

$$W_4^{Haar} = \frac{\sqrt{2}}{2}\begin{bmatrix} 1 & 1 & 0 & 0 \\ 0 & 0 & 1 & 1 \\ 1 & -1 & 0 & 0 \\ 0 & 0 & -1 & 1 \end{bmatrix} \tag{4.36}$$

This computes to that shown in the figure above. The Haar wavelet coefficients, γ, are computed by first inverting the Haar matrix and multiplying by the sample signal vector, $f(t)$:

$$\gamma = \left[W_4^{Haar}\right]^{-1} f(t) \tag{4.37}$$

Suppose that the signal is given by the following series:

$$f(t) = [5|-2|3|1]^T \tag{4.38}$$

A plot of this signal is shown in Fig. 4.22.

The wavelet coefficients are computed to be:

It is possible to cull coefficients on some basis, such as their magnitude with respect to the largest coefficient. We can arbitrarily impose a threshold with respect to the largest absolute value coefficient and remove those coefficients (set to zero) that are at or below this magnitude.

For example, assume an arbitrary threshold of 30%. That is, exclude any wavelet coefficients that are below 30% of the maximum value. The original wavelet coefficients are given by:

$$\gamma = [5.6569|1.4142|-0.7071|-2.1213] \tag{4.39}$$

The largest coefficient in Eq. (4.39) is 5.6569, and 30% of this value is 1.6971. Hence, excluding any coefficients with an absolute value less than this results in the following modified wavelet coefficient vector:

$$\gamma_{30\%} = [5.6569|0|0|-2.1213] \tag{4.40}$$

The signal can be recreated using this modified wavelet coefficient vector, as follows:

$$f(t)_{30\%} = W_4^{Haar}\gamma_{30\%} \tag{4.41}$$

Enumerating Eq. (4.41) and overlaying on the original data results in the plot of Fig. 4.23. Note the comparison between the two signals, indicating some loss of fidelity owing to the removal of the wavelet coefficient. This is a crude representation of the effects of destructive compression on the reconstruction of signals:

The method can be extended easily to any dimension. The use of Haar wavelet transforms allows for modeling signals and signal transients using a subset of data. In effect, this enables compressing larger signals into smaller data sets, or removing transient signal behavior by removing selected coefficients that do not meet *threshold criteria.* The "transient" elements of the original signal are smoothed out and the larger trend of the signal is visible in this plot.

4.5 Process and Policy Implications for Effective Management

Between 2005 and 2008, 566 report of patient deaths throughout the United States related to patient monitoring alarms were received by the FDA and the Manufacturer and User Facility Device Experience (MAUDE) [62]. Many hospital systems have established "Alarm Committees" wherein alarm limit settings, appropriateness of different types of alarms, and methods for mitigating alarms are evaluated, vetted, implemented, and managed. Reporting requirements through The Joint Commission Sentinel Event database can assist hospital systems in identifying what is working or what requirements are relevant to them and as a clearing house for information on best practices, research, and approaches for identifying how to mitigate false, non-

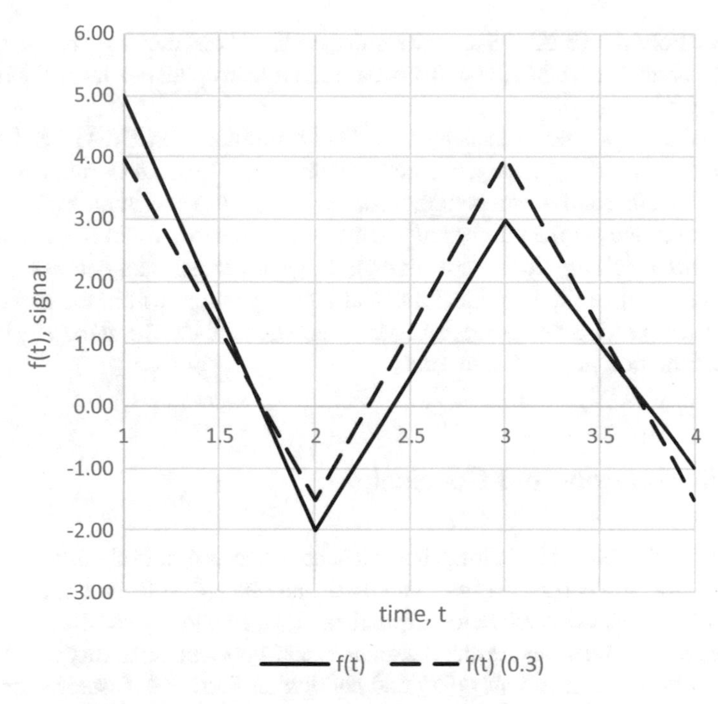

Fig. 4.23 Original signal with recreated signal based on 30% signal threshold

actionable alarms. As alarm system signals are a recognized and tracked patient safety goal, the visibility on the topic is high, even though there does not exist any one-size-fits-all solution approach. Patient safety and risk management has been appearing more often in hospitals, and Quality Officers (usually a clinical staff member, such as a medical doctor or nurse) is charged with the role of managing best practices surrounding improvements in patient safety, tracking events, and providing quality reporting surrounding patient care management.

The Centers for Medicare & Medicaid Services (CMMS) maintains the Hospital Consumer Assessment of Healthcare Providers and Systems (HCAHPS) survey, which "is the first national, standardized publicly presorted survey of patients' perspective of hospital care" [64]. There are several key objectives of this assessment, including obtaining the patient perspective on their care quality. The HCAHPS survey, developed in partnership with the Agency for Healthcare Research and Quality (AHRQ), asks discharged patients 27 questions about their recent hospital stays. Inquiries regarding staff responsiveness, cleanliness, quietness, pain management, medication administration and others are made. Quietness of the environment has a direct bearing on alarm annunciation and noise levels, and the ability to respond to actionable alarms has a direct bearing on patient safety.

Results from the HCAHPS survey results are based on four consecutive quarters of patient surveys, and CMMS publishes participating hospital results on the Hospital Compare web site [64].

Hospital system quality managers are likely familiar with HCAHPS reporting. But, in addition, quality managers together with hospital Alarm Committees should maintain a database of alarm settings, conduct studies of the impact of alarm settings on noise and on the number of alarms issued to care provides, as well as the interventions resulting from these alarms. Together, these data can serve greatly to improve and tune the individual medical device settings within the departments most affected (i.e., intensive care, emergency and general care) and thereby improve patient satisfaction in addition to safety.

4.6 Observations and Conclusions

The use of mathematical techniques in signal and data processing abound in medicine and healthcare, and particularly as relates to the automated collection of data such as from medical devices. Application of mathematical methods to assist in managing and mitigating non-actionable alarm signals is an important capability that manifest in terms of impacts on patient safety and national initiatives to foster better patient experiences within the inpatient hospital setting.

The mathematical treatment outlined in this chapter was intended to illustrate example applications and scope of techniques to processing data in the higher acuity settings of ORs, ICUs and EDs. In each of these locations, chief among the challenges is alarm signal fatigue, which mathematically is a classic Type 1-Type 2 discrimination problem, but carries with it implications on patient safety, efficacy, and clinical responsibility. The use of data from patient populations is important in improving quality and safety associated with the higher acuity patient cohorts found in these wards, and is continuing to become an important adjunct to real-time clinical decision making.

Information on metrics related to alarm counts, HCAHPS and patient experience within the healthcare system are sought by clinical quality managers to provide for continuous process improvement related to patient satisfaction and patient safety. To this end, methods that promote reduced alarm fatigue will have direct impact on patient experience by reducing non-actionable workload on clinicians and facilitating improved patient care management.

What is happening in medicine is a migration from reactive intervention to proactive surveillance. That is, migrating from reacting to the issuance of an alarm signal, such as responding to a patient based on an alarm experienced on a bedside medical device, such as no heart rate (ASYSTOLE) or ventricular tachycardia (VTACH) to identifying that a patient's physiologic parameters indicate a trajectory of decline, such as the onset of respiratory depression. Hence, in terms of future research directions, methods and approaches for mitigating alarm fatigue while increasing the likelihood that when an alarm signal is annunciated that the condition upon which the annunciation is issued is indeed correct. In other words, improved sensitivity (i.e., true positive rate) and specificity (i.e., true negative rate).

The research challenges associated with increasing sensitivity and specificity pertain to multiple factors surrounding data collection, the patient, the environment, the disease condition, training, and staff. Data collection is not uniformly accomplished outside of the intensive care unit setting. Specifically, methods and mechanisms for continuous patient monitoring of various vital signs (e.g.: heart rate, respirations, and other physiological parameters) is normally only uniformly possible inside of the intensive care unit where the workflow is dedicated to one-on-one care and this includes the capture of data in real-time that can be used in conjunction with clinical observations to facilitate rapid decision making on patients. Outside of the intensive care unit setting, continuous monitoring is not generally the norm in many hospitals. Hence, obtaining the requisite data becomes more difficult and unreliable for assisting decision making on measurements. Patient conditions, such as sepsis, respiratory depression, risk of intubation, and ventilator acquired events can reveal themselves in ways that are not entirely uniform across all patients, yet they can be recognized in conjunction with multiple sources of information, including vital signs, bedside observations, and patient presentation. Hence, acquiring and integrating data from multiple sources (i.e., multi-source data fusion) is an area of continuing research interest.

References

1. The Joint Commission, *Sentinel Event Alert, Issue 50, April 8, 2013*, The Joint Commission, (2013)
2. AAMI, Clinical alarms 2011 summit, Association for the Advancement of Medical Instrumentation, (2011)
3. ECRI Institute, *Executive Brief: Top 10 Health Technology Hazards for 2017*, ECRI (2016)
4. The Joint Commission, *R3 Report I: Alarm system safety; Issue 5, December 11, 2013*, The Joint Commission, (2013)
5. Gartner, *Gartner IT Glossary Definition for Big Data*, 18 Oct 2016. (Online). Available: http://www.gartner.com/it-glossary/big-data
6. S.E. White, A review of big data in health care: challenges and opportunities, *Open Access Bioinformatics*, 31 Oct 2014
7. M. McNickle, 5 reasons medical device data is vital to the success of EHRs, *Healthcare IT News,* 5 Jan 2012
8. Bernoulli Enterprises, Inc., *Bernoulli Health*, (Online). Available: http://bernoullihealth.com/solutions/medical-device-integration/. Accessed Nov 21 2016
9. Qualcomm, (Online). Available: https://www.qualcomm.com/solutions/health-care/hospital-connectivity-and-integration. Accessed Nov 21 2016
10. J.R. Zaleski, Semantic data alignment of medical devices supports improved interoperability. Med. Res. Arch. **4**(4) (2016)
11. J.R. Zaleski, *Connected Medical Devices: Integrating Patient Care Data in Healthcare Systems* (Health Information Management Systems Society, Chicago, 2015), pp. 7–56
12. J.R. Zaleski, *Integrating Device Data into the Electronic Medical Record: A Developer's Guide to Design and a Practitioner's Guide to Application* (Publicis Publishing, Erlangen, 2009), pp. 39–69

13. Z. Obermeyer, E.J. Emanuel, Predicting the future-big data, machine learning, and clinical medicine. N. Engl. J. Med. **375**(13), 29 (2016)
14. J.R. Zaleski, *Medical Device Data and Modeling for Clinical Decision Making* (Artech House, Boston, 2011), pp. 173–178
15. M. Field, K. Lohr, *Guidelines for Clinical Practice: From Development to Use* (Institute of Medicine, National Academy Press, Washington, 1992)
16. National Guidelines Clearinghouse, *Inclusion Criteria*, 1 June 2014. (Online). Available: https://www.guideline.gov/help-and-about/summaries/inclusion-criteria. Accessed Nov 17 2016
17. Institute for Clinical Systems Improvement, *Health Care Protocol: Rapid Response Team*, 4th edn. 7 2011. (Online). Available: https://www.icsi.org/_asset/8snj28/RRT.pdf. Accessed Nov 23 2016
18. J. Li, *University of Florida Course Notes: EEL 6537—Introduction to Spectral Analysis—Spring 2010*, 2010. (Online). Available: http://www.sal.ufl.edu/eel6537_2010/LSP.pdf. Accessed Nov 21 2016
19. J. Long, *Recovering Signals from Unevenly Sampled Data*, 24 11 2014. (Online). Available: http://joseph-long.com/writing/recovering-signals-from-unevenly-sampled-data/. Accessed Nov 21 2016
20. A. Mathias, F. Grond, R. Guardans, D. Seese, M. Canela, H.H. Diebner, Algorithms for spectral analysis of irregularly sampled time series. J. Stat. Softw. **11**(2), 5 (2004)
21. T. Ruf, The Lomb-Scargle periodogram in biological rhythm research: analysis of incomplete and unequally spaced time-series. Biol. Rhythm Res. **30**(2), 178–201 (1999)
22. J. Pucik, *Heart Rate Variability Spectrum: Physiologic Aliasing and Non-Stationarity Considerations*, Bratislava (2009)
23. "Lomb Scargle periodogram for unevenly sampled time series," 10 Jan 2013. (Online). Available: https://www.r-bloggers.com/lomb-scargle-periodogram-for-unevenly-sampled-time-series-2/. Accessed 15 Nov 2016
24. G.L. Brethorst, *Frequency Estimation and Generalized Lomb-Scargle Periodograms*, 20 Apr 2015. (Online). Available: http://bayes.wustl.edu/glb/Lomb1.pdf
25. J.R. Zaleski, *Investigating the Use of the Lomb-Scargle Periodogram for Heart Rate Variability Quantification*, April 2015. (Online). Available: http://www.medicinfotech.com/2015/04/lombscargle-periodogram-measure-heart-rate-variability
26. M. Malik, Heart rate variability: standards of measurement, physiologic interpretation, and clinical use, in *Task Force of the European Society of Cardiology: the North American Society of Pacing Electrophysiology* (1996)
27. M.M. Corrales, B. de la Cruz Torres, A.G. Esquival, M.A.G. Salazar, J.N. Orellana, Normal values of heart rate variability at rest in a young, healthy and active Mexican population. SciRes **4**(7), 377–385 (2012)
28. M. Saeed, M. Villarroel, A. Reisner, G. Clifford, L. Lehman, G. Moody, T. Heldt, T. Kyaw, B. Moody, R. Mark, Multiparameter intelligent monitoring in intensive care II (MIMIC-II): a public access ICU database. Crit. Care Med. **39**(5):952–960 (2011)
29. Polar, *Heart Rate Variability*, 23 Apr 2015. (Online). Available: http://support.polar.com/us-en/support/Heart_Rate_Variability__HRV_. Accessed 15 Nov 2016
30. K.C. Bilchick, R.D. Berger, Heart rate variability. J. Cardiovasc. Electrophysiol. **17**(6), 691–694 (2006)
31. "Physiotools," 23 Apr 2015. (Online). Available: http://physionet.org/physiotools/ecgsyn/. Accessed 23-Apr 2015
32. P. Stoica, J. Li, H. He, Spectral analysis of nonuniformly sampled data: a new approach versus the periodogram. IEEE Trans. Sig. Process. **57**(3), 843–858 (2009)
33. M. Zechmeister, M. Kuerster, The generalised Lomb-Scargle periodogram—A new formalism for the floating-mean and Keplerian periodograms. Astron. Astrophys. **20**, 1 (2009)
34. S. Ahmad, T. Ramsay, L. Huebsch, S. Flanagan, S. McDiarmid, I. Batkin, L. McIntyre, S.R. Sundaresan, D.E. Maziak, F.M. Shamji, P. Hebert, D. Fergusson, A. Tinmouth, A.J. Seely, Continuous multi-parameter heart rate variability analysis heralds onset of sepsis in adults. PLoS ONE **4**(8) (2009)

35. N. Stevens, A.R. Giannareas, V. Kern, A.V. Trevino, M. Fortino-Mullen, Smart alarms: multivariate medical alarm integration for post CABG surgery, in *ACM SIGHIT International Health Informatics Symposium (IHI 2012)*, Miami, FL (2012)
36. V. Herasevich, S. Chandra, D. Agarwal, A. Hanson, J.C. Farmer, B.W. Pickering, O. Gajic, V. Herasevich, The use of an electronic medical record based automatic claculation tool to quantify risk of unplanned readmission to the intensive care unit: a validation study. J. Crit. Care **26** (2011)
37. O. Gajic, M. Malinchoc, T.B. Comfere, M.R. Harris, A. Achouiti, M. Yilmaz, M.J. Schultz, R.D. Hubmayr, B. Afessa, J.C. Farmer, The stability and workload index for transfer score predicts unplanned intensive care unit patient readmission: Initial development and validation. Crit. Care Med. **36**(3) (2008)
38. U.R. Ofoma, S. Chandra, R. Kashyap, V. Herasevich, A. Ahmed, O. Gajic, B.W. Pickering, C.J. Farmer, Findings from the implementation of a validated readmission predictive tool in the discharge workflow of a medical intensive care unit. AnnalsATS **11**(5) (2014)
39. V.M. Boniatti, M.M. Boniatti, C.F. Andrade, C.C. Zigiotto, P. Kaminski, S.P. Gomes, R. Lippert, M.C. Diego, E.A. Felix, The modified integrative weaning index as a predictor of extubation failure. Respir. Care **59**(7) (2014)
40. J. Garah, O.E. Adiv, I. Rosen, R. Shaoul, The value of Integrated Pulmonary Index (IPI) during endoscopies in children. J. Clin. Monit. Comput. **29**, 773–778 (2015)
41. M.B. Weinger, L.A. Lee, No patient shall be harmed by opioid-induced respiratory depression. APSF Newslett. **26**(2), 21–40 (2011)
42. J.F. Mudge, L.F. Baker, C.B. Edge, J.E. Houlahan, Setting an optimal alpha that minimizes errors in null hypothesis significance tests. PLoS ONE **7**(2) (2012)
43. M.A. Cretikos, R. Bellomo, K. Hillman, J. Chen, S. Finfer, A. Flabouris, Respiratory rate: the neglected vital sign. Med. J. Aust. **188**(11), 657–659 (2008)
44. B. Page, "Capnography helps conscious patients too, J. Emerg. Med. Serv. **29** (2010)
45. J.R. Zaleski, *Mathematical Techniques for Mitigating Alarm Fatigue*, 14 Oct 2014. (Online). Available: http://www.medicinfotech.com/2014/10/mathematical-techniques-mitigating-alarm-fatigue. Accessed 15 Nov 2016
46. I. MacDonald, Hospitals rank alarm fatigue as top patient safety concern. Fierce Healthcare, 22 Jan 2014
47. G. Gachco, J. Perez-Calle, A. Barbado, J. Lledo, R. Ojea, V. Fernandez-Rodriguez, Capnography is superior to pulse oximetry for the detection of respiratory depression during colonoscopy. Rev. Esp. Enferm. Dig. **102**(2), 86–89 (2010)
48. R.G. Soto, E.S. Fu, J.H. Vila, R.V. Miquel, Capnography accurately detects apnea during monitoried anesthesia care. Anesth. Analg. **99**, 379–382 (2004)
49. D. Carr, A. Cartwright, Rationales and applications for capnography monitoring during sedation. Clinical Foundations (2011)
50. B.S. Kodali, Capnography outside the operating rooms. Anesthesiology **118**(1), 192–201 (2013)
51. M. Wong, A. Mabuyi, B. Gonzalez, First national survey of patient-controlled analgesia practices, A Promise to Amanda Foundation and the Physician-Patient Alliance for Health & Safety, March-April 2013
52. B. Sullivan, 5 things to know about capnography and respiratory distress, *EMS1 News,* 15 Oct 2015
53. G. Welch, G. Bishop, *An Introduction to the Kalman Filter,* (Chapel Hill, NC, 27599-3175: ACM, Inc., pp. 24–25) (2001)
54. L. Kleeman, *Understanding and Applying Kalman Filtering*, (Online). Available: http://biorobotics.ri.cmu.edu/papers/sbp_papers/integrated3/kleeman_kalman_basics.pdf. Accessed 15 Nov 2016
55. R. Sameni, M. Shamsollahi, C. Jutten, Filtering electrocardiogram signals using the extended Kalman Filter, in *27th Annual International Conference of the IEEE Engineering in Medicine and Biology Society (EMBS), Sep 2005, Shanghai, China*, Shanghai, China, 2005

56. R.F. Suoto, J.Y. Ishihara, A robust extended Kalman Filter for discrete-time systems with uncertain dynamics, measurements and correlated noise, in *American Control Conference*, St. Louis, MO, 2009
57. R. Gao, R. Yan, Chapter 2: from fourier transform to wavelet transform: a historical perspective, *Wavelets: Theory and Applicationf for Manufacturing* (LLC, Springer Science + Business Media, 2011), p. 18
58. B.D. Patil, *Introduction to Wavelet*, Spring 2014. (Online). Available: https://inst.eecs.berkeley.edu/~ee225b/sp14/lectures/shorterm.pdf. Accessed 13 Dec 2016
59. G. Dallas, Wavelets *4 Dummies: Signal Processing, Fourier Transforms and Heisenberg*, 14 5 2014. (Online). Available: https://georgemdallas.wordpress.com/2014/05/14/wavelets-4-dummies-signal-processing-fourier-transforms-and-heisenberg/. Accessed 13 Dec 2016
60. C.S. Burrus, R.A. Gopinath, H. Gao, *Introduction to Wavelets and Wavelet Transforms: A Primer* (Simon & Schuster, Upper Saddle River, 1998)
61. G. Strang, Wavelet transforms versus fourier transforms. Bull. Am. Math. Soc. **28**(2), 288–305 (1993)
62. A.C. Bridi, T.Q. Louro, R.C. Lyra da Silva, Clinical Alarm in inensive care: implications of alarm fatigue for the safety of patients, Rev. Lat. Am. Enfermagem **22**(6), 1034–1040, Nov–Dec (2014)
63. CMS, *HCAHPS: Patient's Perspectives of Care Survey*, (Online). Available: https://www.cms.gov/Medicare/Quality-Initiatives-Patient-Assessment-Instruments/HospitalQualityInits/HospitalHCAHPS.html. Accessed 31 August 2017
64. CMS, (Online). Available: www.hospitalcompare.hhs.gov

John Zaleski, Ph.D., CAP, CPHIMS is Chief Analytics Officer of Bernoulli Enterprise, Inc. (http://bernoullihealth.com). Dr. Zaleski brings 22 years of experience in researching and ushering to market products to improve healthcare. He received his Ph.D. from the University of Pennsylvania, with a dissertation that describes a novel approach for modeling and prediction of post-operative respiratory behavior in post-surgical cardiac patients. Dr. Zaleski has a particular expertise in developing and implementing point-of-care applications for hospital enterprises. Dr. Zaleski is the named inventor or co-inventor on eight issued U.S. patents related to medical device interoperability and clinical informatics. He is the author of peer-reviewed articles and conference papers on clinical use of medical device data, and wrote three seminal books on integrating medical device data into electronic health records and the use of medical device data for clinical decision making, including the #1 best seller of HIMSS 2015 on connected medical devices.

Chapter 5
A Novel Big Data-Enabled Approach, Individualizing and Optimizing Brain Disorder Rehabilitation

Marketa Janatova, Miroslav Uller, Olga Stepankova, Peter Brezany and Marek Lenart

Abstract Brain disorders occur when our brain is damaged or negatively influenced by injury, surgery, or health conditions. This chapter shows how the combination of novel biofeedback-based treatments producing large data sets with Big Data and Cloud-Dew Computing paradigms can contribute to the greater good of patients in the context of rehabilitation of balance disorders, a significant category of brain damage impairments. The underlying hypothesis of the presented original research approach is that detailed monitoring and continuous analysis of patient´s physiological data integrated with data captured from other sources helps to optimize the therapy w.r.t. the current needs of the patient, improves the efficiency of the therapeutic process, and prevents patient overstressing during the therapy. In the proposed application model, training built upon two systems, Homebalance—a system enabling balance training and Scope—a system collecting physiological data, is provided both in collaborating

M. Janatova
Joint Department of Biomedical Engineering, Department of Rehabilitation Medicine, First Faculty of Medicine of CU and General Teaching Hospital in Prague, CTU and Charles University (CU), Prague, Czech Republic
e-mail: marketa.janatova2@lf1.cuni.cz

M. Uller
Robotics and Machine Perception Department, CIIRC CTU, Prague, Czech Republic
e-mail: Miroslav.Uller@cvut.cz

O. Stepankova
Biomedical Engineering and Assistive Technologies Department, CIIRC CTU, Prague, Czech Republic
e-mail: olga.stepankova@cvut.cz

P. Brezany (✉) · M. Lenart
Faculty of Computer Science, University of Vienna, Vienna, Austria
e-mail: Peter.Brezany@univie.ac.at

M. Lenart
e-mail: Marek.Lenart@univie.ac.at

P. Brezany
SIX Research Group, Brno University of Technology, Brno, Czech Republic

A. Emrouznejad and V. Charles (eds.), *Big Data for the Greater Good*,
Studies in Big Data 42, https://doi.org/10.1007/978-3-319-93061-9_5

rehabilitation centers and at patient homes. The preliminary results are documented using a case study confirming that the approach offers a viable way towards the greater good of a patient.

Keywords Brain damage · Biofeedback · Data analysis · Dew computing

5.1 Introduction

Brain disorders can negatively influence function of most parts of human body. Necessary medical care and rehabilitation is often impossible without considering the state-of-the-art neuroscience research achievements and close cooperation of several diverse medical, care providing, IT and other specialists who must work jointly to choose methods that can improve and support healing processes as well as to suggest new treatment processes based on better understanding of underlying principles. The key to their treatment decisions are data resulting from careful observation or examination of the patient during his/her exercise ensured as a part of the long-term therapy (which can last months or even years) and information/knowledge extracted from these data. Because of the impressive advances in scientific domains, namely, the neuroscience, as well as in technology that allows e.g., to ensure on-line monitoring of patient's activity during exercise through streaming data from body area network of sensors, the volume and complexity of the captured data have been continuously increasing. The resulting data volumes are now reaching the challenges of the Big data technology [18] with respect to the four Vs used to characterize this concept: volume (scale or quantity of data), velocity (speed and analysis of real-time data), variety (different forms of data, often from different data sources), and veracity (quality assurance of the data). We believe that the achievements of Big Data in combination with the novel Cloud-Dew Computing [29, 30] paradigm could introduce the development of a new generation of non-invasive brain disorder rehabilitation processes and, in such a way, contribute to the greater good of patients thanks to higher efficiency of rehabilitation process and better quality of life as confirmed by a case study briefly mentioned in Sect. 5.5 of this chapter [20]. These issues are addressed by our project, the results of which are presented here. In the rest of this section, we first provide the background-information as a motivation for our work, discuss the relevant state of the art, introduce our long-term research objectives and, finally, outline the structure of the chapter.

5.1.1 *Background Information and the State of the Art*

The use of modern technologies for the rehabilitation of patients with brain damage has made a remarkable progress in the past decade. The therapists are trying to incorporate into their therapy sessions various tools originally designed for entertainment, e.g., combine therapy with computer games or with virtual reality, to improve attractiveness of the traditional beneficial exercise and adherence of the patients to the therapy plan. Such an enrichment represents a significant addition to conventional procedures, whose experience and best practices have steered the development of new therapeutic technologies incorporating the latest achievements of computational intelligence, sensing technique, virtual reality, video games, and telemedicine. While currently the needed therapy (some form of training) is primarily provided in rehabilitation centers; our vision is to make it available even in patients' homes, because home therapy, without direct participation of a therapist, can save the patient discomfort of traveling to centers, makes the rehabilitation more attractive, provides motivation to exercise, and so, it can be more effective. Further, home therapy combined with telehealth solutions could target much bigger groups of well selected patients without compromising quality of care. All this contributes to the greater good of the considered individual patients and of efficient utilization of resources in healthcare system. Resulting big data, locally collected in patients' home and transmitted to the rehabilitation center, where they are first integrated (and significantly enlarged) with the data collected in the center, processed and analyzed can further provide precious source that can help us in understanding the function of human brain.

Balance disorder of the central and peripheral system is a significant symptom that causes functional deficit in common activities. Patients after brain damage often face situations (e.g. using escalators, orientation in city traffic) that can be possibly dangerous or even impossible to manage. Consequently, balance training is one of the main focuses of brain disorder rehabilitation. Most modern balance training environments are interactive and built up on measuring instruments called force platforms (or force plates) that measure the ground reaction forces (their summary vector is often called as center of pressure) generated by a body standing on or moving across them, to quantify balance, gait and other parameters of biomechanics [11]. Methods based on virtual reality technology and biological feedback are appropriately supplementing conventional rehabilitation procedures. In this context, in our lab, we have developed a stabilometric system [2, 3, 15], based on the WiiFit balance board (for details see http://www.wiifit.com/), with a set of 2D or 3D training scenes so that patients can experience situations appearing in these scenes without any risk. Commonly used 2D gaming systems are not suitable for this purpose. The therapy includes active repetitive game-like trainings. The patient standing on the force platform is set the task to control his/her center of pressure (COP) movements[1] to achieve goals of the games. The COP movement is visualized in the virtual environment to provide the visual feedback. The difficulty of trainings and sensitivity of sensors can be adjusted with respect to the current patient's state. Patient's state is objectively evaluated in the beginning and during the entire rehabilitation process. The efficiency of patient's COP movement is measured and presented as a score value. Data is analyzed in both time and frequency domain. Graphs and raw values can be exported. During the ther-

[1]The corresponding trajectory is called the statokinesiogram; it is the map of the COP in the anteroposterior direction versus the COP in the mediolateral direction [25].

apy is biological feedback established by means of the visualization of the projection of position of the center of gravity, acoustic signal, or electro tactile stimulation of tongue [5].

Similar objectives have been followed by other researchers Esculier et al. [6], Cho et al. [9], Gatica-Rojas and Mendez-Rebolledo [12], Giggins et al. [13], Gil-Gomez et al. [14], Borghese et al. [4]. Papers [13, 14] analyze the state of the art in balance rehabilitation systems supported by the virtual reality and biofeedback technologies, respectively. The WiiFit force platform based rehabilitation systems and experience gained from the real application environments are discussed in Esculier [6], Cho et al. [9], Gatica-Rojas and Mendez-Rebolledo [12], Borghese et al. [4]. A game engine that involves advance characteristics needed to support rehabilitation at home is described in Bakker et al. [1]; here, a rule-based component partially steers the therapeutic process.

Dew computing is a choice for the implementation of the above application processes in a distributed way. It is a very new computing paradigm that sets up on top of Cloud computing and overcomes some of its restrictions like the dependency on the Internet. There are different definitions of Dew computing [24, 29, 30]. In our effort, we follow Y. Wang's definition of Dew computing: "*Dew computing is an on-premises computer software-hardware organization paradigm in the cloud computing environment where the on-premises computer provides functionality that is independent of cloud services and is also collaborative with cloud services. The goal of dew computing is to fully realize the potentials of on-premises computers and cloud services*" [31]. This guarantees that the offered services are independent of the availability of a functioning Internet connection. The underlying principle for this is a tight collaboration between on-premise and off-premise services based on an automatic data exchange among the involved compute resources.

5.1.2 Research Objectives and Organization of the Chapter

Investigations in the above discussed research showed that regular training has a beneficial effect especially on the balance, motor skills, spatial orientation, reaction time, memory, attention, confidence and mental well-being of the user. However, the personalized application of the available systems for therapeutic processes requires careful choice of appropriate settings for large number of system parameters, values of which cannot be estimated purely from few simple measurements (e.g. weight and height of the patient). The system parameters must be modified w.r.t. to current patient´s state as well as to the set of multiple diseases the patient is suffering from. This is the first reason why a large search space should be considered when planning the type of therapeutic processes to be recommended for system parameter settings. The second reason is related to the fact that individual combination of therapeutic processes is recommended for each patient and this combination influences the parameter values for individual processes. All this leads to combinatorial explosion that must be faced. Our objective is to contribute to the solution of these challenges

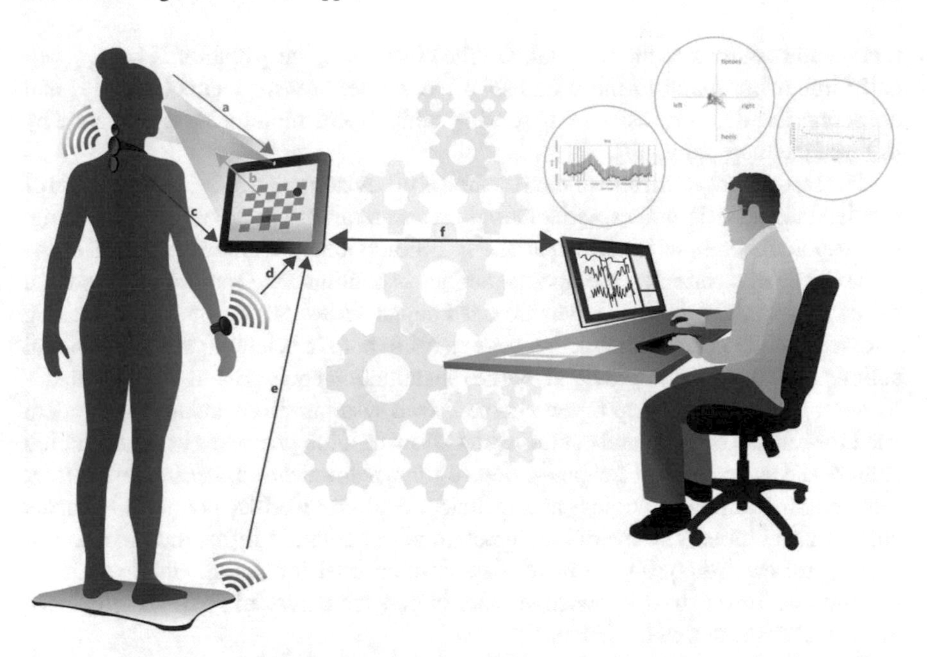

Fig. 5.1 Balance training supported by biofeedback data: (a) patient face is observed by a camera (a) post-analysis can identify e.g., specific emotions, concentration, etc.); (b) patient gets audiovisual feedback; (c) sensing neck muscle tension; (d) measuring pulse frequency; (e) data streams from balance force platform sensors; (f) data transfer/processing between patient and therapist areas

by an approach that includes biofeedback-based methods. Figure 5.1 illustrates basic elements of biofeedback realization. The system parameter setting currently relies mainly on subjective observation of the status and behavior of the patient by the persons steering the therapy—this type of human expert supervision is time consuming, sub-optimal, and unproductive. The challenge and an ambitious research goal is to provide technical support for design of an optimal therapeutic process for each individual patient. We suggest applying for that purpose methods of machine learning or case-based reasoning. A prerequisite for this approach is to collect enough training data that fully characterize the detailed advance of the patient during the rehabilitation processes and his/her therapies.

An interesting issue is also the choice of the optimal set of sensors for a given patient. In theory, it seems that the more data we can collect from the considered patient, the better. On the other hand, one should not forget that placement of each BAN (body area network) sensor makes the preparation for the exercise more demanding and causes additional discomfort for the patient who tries to proceed with therapy at home. That is why special attention should be given to the selection of those sensors that provide necessary information and the patient has no problems to apply them on. Obviously, the set of sensors for home therapy cannot be the same for all patients under these conditions. The appropriate choice can be significantly supported by careful analysis of patient´s data collected during his/her therapeutic sessions in the hospital. On the other hand, the situation is being gradually improving thank to the progress in the wireless sensor technology [19] and non-contact and non-invasive body sensing methods, e.g. monitoring temperature by infrared thermography [17].

The second long-term aim of our research is to develop an adaptive optimization of involved therapeutic processes including precise monitoring, assessment, learning, and response cycles and based on patient-specific modeling. The current patient status achieved by goal-oriented therapeutic actions is continuously monitored through a set of sensors. The received synchronized signal series can be interpreted taking into account the current knowledge base status to help in selecting the next optimal actions from the rich set provided by the rehabilitation environment to be applied.

Our approach is built up on two home-grown systems: Homebalance—a system used for balance training and capturing data from training processes in rehabilitation centers and at homes and Scope—a device for synchronization, storing and further transmitting various biological and technical data provided by connected sensors and technical tools. The overall architecture of our solution infrastructure is based on Cloud-Dew Computing, a novel paradigm optimal for the application pattern we address. To our best knowledge there is no such a system addressing balance disorders rehabilitation available.

The rest of the chapter is organized as follows. Section 5.2 outlines the overall patient's treatment workflow and associated data flows. Capturing data in our rehabilitation approach is discussed in Sect. 5.3. These raw data are processed, analyzed and visualized as described in Sect. 5.4. What can be learned from the analysis results and how they improve the rehabilitation processes? Section 5.5 deals with this topic. The overall system architecture including elements of cloud and dew computing is presented in Sect. 5.6. Finally, we conclude and outline the future research in Sect. 5.7.

5.2 Treatment Workflow and Associated Data Space Development

Prior to entering the rehabilitation phase, we focus on, there is certain patient's development trajectory (pre-rehabilitation → rehabilitation → post-rehabilitation) associated with data collection incrementally creating a big Data Space, as depicted in Fig. 5.2. For example, in the Brain damage prehistory phase, an Electronic Health Record of the patient has been created and gradually developed. The concept of Scientific Data Space has been originally introduced by Elsayed [8]. This chapter focuses on the Rehabilitation processes conducting the development of the Rehabilitation Data Sub-Space (RDSS). For specific data analysis purposes, RDSS can be flexible enlarged by integration with other data space elements. RDSS includes three types of data:

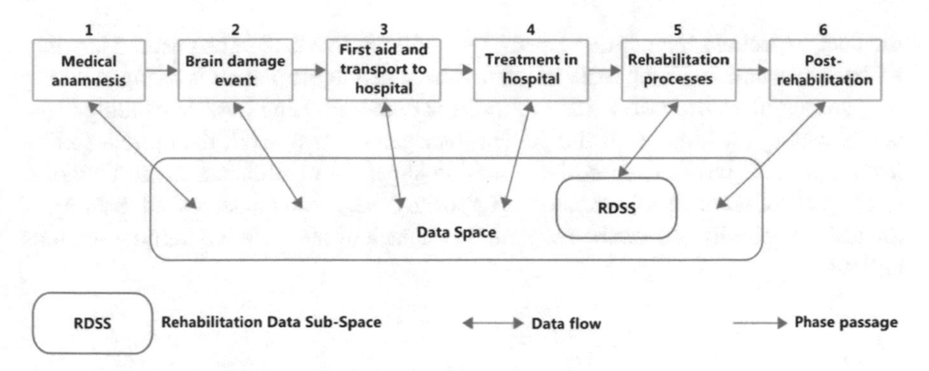

Fig. 5.2 Treatment workflow and data space development

- Primary data—data captured during therapies (e.g., data produced by sensors and balance force platform, game event data, videos, etc.)
- Derived data—including statistical and data mining models, results of signal processing, images resulted from the visualization processes, etc.
- Background data—providing description of applied workflows, reports on scientific studies, relevant publications, etc.

The initial session and patient status estimation is the first step in the Rehabilitation phase of the treatment trajectory. Here, the Berg Balance Scale [17] is typically applied. It is a 14-item objective measure designed to assess static balance and fall risk in adult populations, with the following interpretation: 0–20, wheelchair bound; 21–40, walking with assistance; 41–56, independent. Mining data subspace constructed from data elements produced by Phases 1–4 of Fig. 5.2 can reveal additional knowledge helpful for the assessment.

5.3 Data Capturing

In this section, the real data gathered during the sample experiments (Homebalance training sessions) will be explored in more detail. The relevance of individual data streams will be considered and additional data that may be useful during further analyses will be proposed.

The typical training (rehabilitation) session using the Homebalance system and the Scope device used for capturing biofeedback data as described in the previous section consists of three consecutive tasks that the probands[2] are to perform: a simple chessboard scene and two scenes in 3D virtual environments.

Each of the considered training sessions has several levels of complexity given by the used parameters of the used task, e.g. maximal time allowed for accomplishing

[2]In different medical fields, the term proband is often used to denote a particular subject, e.g., person or animal, being studied or reported on.

the task, expected scope of movement, type of cognitive task to be ensured together with the movement. Complexity of the task assigned to the patient is changed during the exercise in accordance to the biofeedback data informing how demanding is the actual setting for him/her. If the patient manages the task well, the current setting level can be slightly increased by a session supervisor (clinician, home therapist, intelligent software agent, etc.). In the opposite case, it can be decreased. Scaling of the task complexity is described separately for each of the included training sessions bellow.

5.3.1 *The "Chessboard Scene"*

In the chessboard scene (Fig. 5.3a), the proband is shown a chessboard with a sphere and the target location on the chessboard is highlighted. Standing on the balance force platform, the proband is instructed to move the sphere by shifting his/her center of gravity (COG). A body's COG is the point around which the resultant torque due to gravity forces vanishes. If the projection of COG to the platform is shifting in the correct direction and remains there for the requested time, the sphere moves to the highlighted location. The process is then repeated several times, with different target locations.

Fig. 5.3 Examples of virtual scenes: **a** the chessboard scene; **b** the "city street scene"; **c** the "water maze" virtual scene-starting position (beginning of the scene); **d** end position: the patient has to choose the right combination of colors and shapes

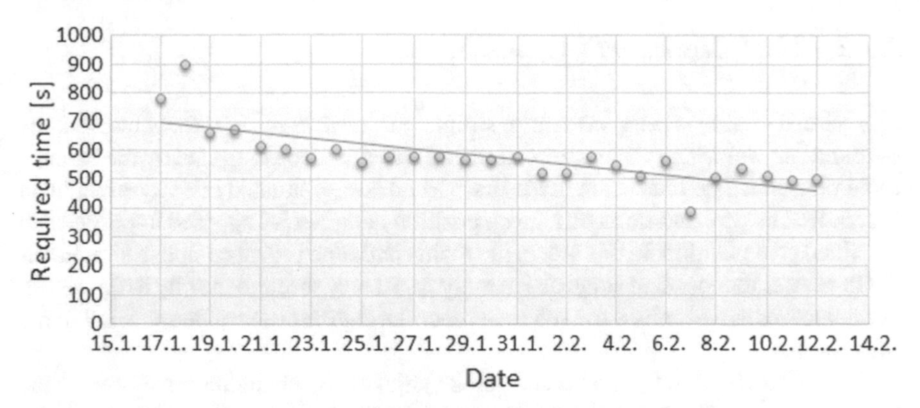

Fig. 5.4 Development of the training performance of a patient during the period of 4 weeks

The difficulty of training session can be adjusted to the current patient's state. Tasks used for the first phase of rehabilitation process are designed with target locations near the middle of the chessboard, with short time of quiet stance on each required position. If the patient completes the basic task correctly, the therapist chooses training scene with target locations in outlying positions and longer time of quiet stance on each position.

Figure 5.4 shows changes of time a patient requires to complete the same reference scene during one-month therapy. This patient responds well to the therapy and her/his results are steadily improving. Even in this positive case, one cannot fully rely on improvement in the longer time interval; in the example graph, it is, e.g., shown that the result of the second day (18.1.) is significantly worse than that of the first day (17.1.). This might be caused by the fact that while the first day the exercise has been done under supervision of a therapist, the second day the patient has been on his own what caused some stress as can be verified from the BAN (body area network) data stored by the Scope device. The parameter setting must be adjusted to this change on the fly to make sure that the patient remains confident that he/she can accomplish what he/she is asked for. This helps him/her to build self-confidence in his/her abilities and motivates him/her to proceed with the designed training process. The system can learn to do the appropriate necessary adjustment (semi-)automatically using the results of the analysis of the collected data (see the subsequent parts 3.4 and 4)—in this way, the system is becoming smarter, more efficient, and more user friendly. Moreover, it leads to discovery of parameters and attitudes evoking more optimism and motivation for the therapy (even self-therapy after the end of the clinician supervising). The impact of it is illustrated, e.g., on 19.1. showing improved training performance.

5.3.2 The "City Street" Virtual Scene

The virtual reality scenes present a simple 3D environment through which the proband can walk. The walking direction and speed is also controlled by the balance force platform. The screenshot of the first virtual scene, situated in a city street, is in Fig. 5.3b. The proband starts on a given position and then is supposed to progress to a defined goal position. In this scene, the main challenge is to accomplish simultaneously several independent subgoals, namely find a way through the city to the target, not get lost or run over by a car and remember during the task a code specified in the start location.

The difficulty of the virtual scene can be adjusted by changing sensitivity of the balance force platform, changing distance of the final target place and customizing the number and parameters of obstacles (e.g., the number and speed of cars). The scene can be controlled by keyboard or joystick during initial phase of rehabilitation process, when the patient is getting acquainted with the new virtual environment before he/she is asked to control movements through the scene by shifting of his/her center of gravity on the balance force platform.

5.3.3 The "Water Maze" Virtual Scene

The second scene represents a maze of narrow walkways above a body of water (Fig. 5.3c, d). The proband has quite clear overview of the scene and the goal position, but the paths he/she should walk on are very narrow, so it is relatively challenging (when using the WiiFit balance board as controller) to keep the straight movement and not falling into water below.

The patient has to remember a code (combination of colors and shapes at the beginning of the scene). He/she should select the correct combination after he/she reaches final position. The difficulty of the task can be adjusted by changing the number of objects the patient should remember.

The duration of the chessboard and virtual reality scenes typically ranges from 2 min in chessboard scene up to 5 min in 3D virtual scenes; naturally, the durations depend not only on scene complexity but also on the skill and the spatial orientation ability of the proband.

5.3.4 Description of the Collected Physiological Data

In the training experiments performed so far, the following physiological data were collected: ECG—electrocardiogram; EIMP - electrical impedance of the chest, a variant of electro impedance plethysmography measured by the same electrodes as those used for measuring ECG; GSR—galvanic skin response, measured by sensors attached to the fingers of the proband; ACCX, ACCY, ACCZ—accelerometers placed on the back side of proband's neck, for detecting the movement of the proband; and TEMP—the body temperature, the sensor placed on the back side of proband's neck. All the data also contains a time information (UTC timestamp provided during synchronization and storing data by the Scope device) and are sampled with sampling frequency roughly 1 kHz (the elapsed time between two consecutive samples is approximately 1 ms).

An example of the raw data gathered during one task of a therapy session is shown in Fig. 5.5, where one can watch and compare the values of the signals measured simultaneously, namely ECG, EIMP, GSR, TEMP and the magnitude of the sum of vectors ACCX and ACCY, which is denoted as ACC (XY module).

Most of these data are not preprocessed—the only exception is the data from accelerometers that were demeaned; the data ranges correspond to the ranges of the sensors used. The red lines mark specific observations provided by the therapist who has been present during the considered session. In the addressed case, these observations identify the signs of patient's discomfort the therapist can feel. In some other case, these additional signs can be used to identify occurrence of unexpected events (anomalies) from outside the simulated scenario (e.g. sensor adjustment, outside disturbance) that might have affected the proband's performance.

Let us try to review how the therapist interprets the data in Fig. 5.5. During the first few seconds he/she visually checks the sensor connection by verifying that the measured values are reliable (their scope and changes correspond to their expected image) and concludes that all sensors are working well. Here all the sensors behave according to expectations and the session can proceed. From now on he/she can fully concentrate on observation of patient's performance and denote the moments when he/she feels some tension in patient's behavior. Generally, low levels of GSR are a clear sign of stress. This is confirmed at the patient in the left column in the moment of the first red line (30 s after the start approximately). On the other hand, the sign of stress occurs in GSR signal at about 120 s after the start but it is not identified by the therapist who notices something unusual slightly later at the time point 145 s, where a peak occurs in the EIMP signal. While the last 3 red lines, in time points 145, 240 and 270 s, correspond to the situation when the high peak occurs in EIMP and simultaneously low peak appears in GSR this is not the case of the low GSR peak in the time point 120 s. Indeed, this patient often showed a visible sign of relief which was demonstrated by a deep breath he took when he accomplished one of the assigned subgoals. This points to the individual behavior pattern related to stress that is characteristic to this specific patient. This observation can be reused later during individual training sessions without direct supervision of the therapist. On the other hand, the low peak of GSR in the time 120 s shows that one cannot fully rely on the deep breath to estimate occurrence of stress with the considered patient.

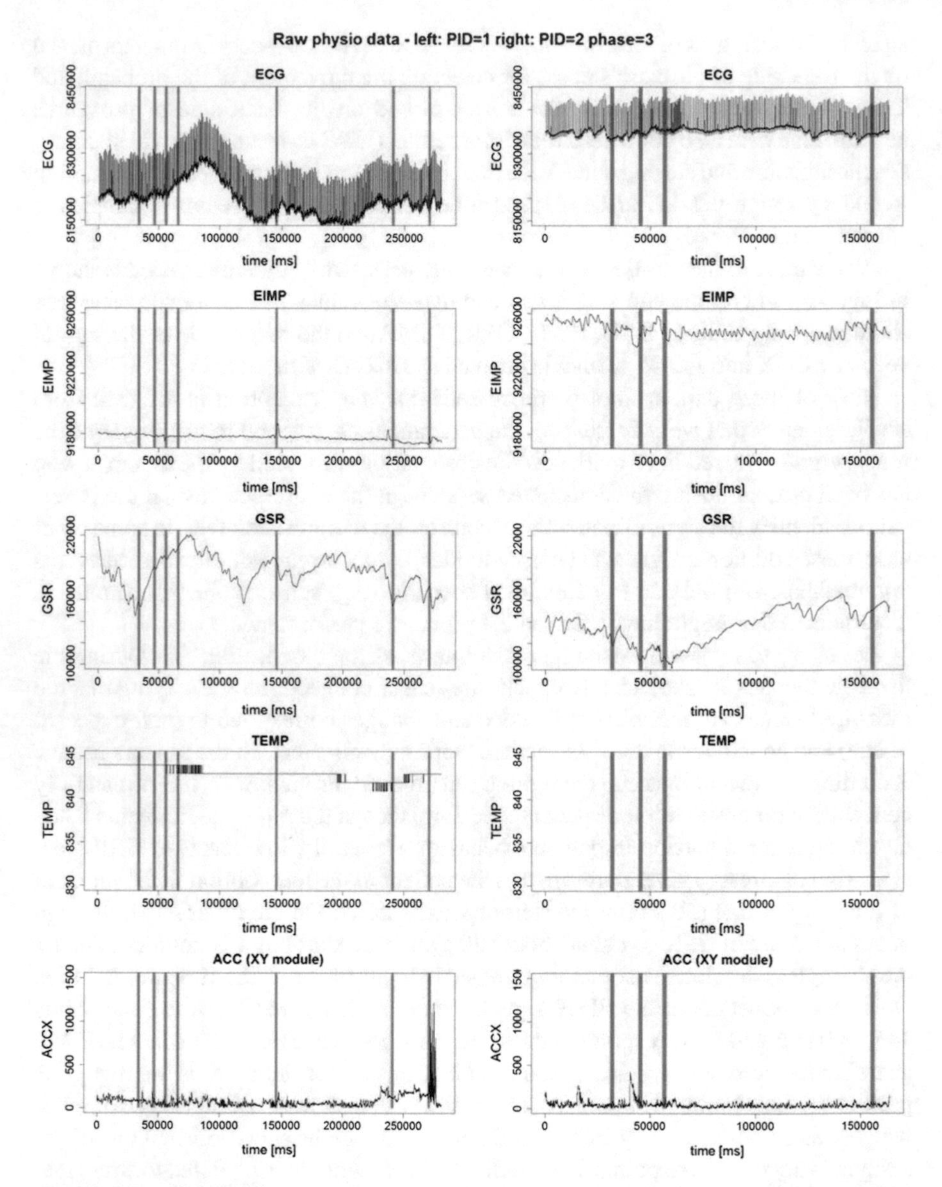

Fig. 5.5 Two columns visualize collected raw data for two probands training on the same virtual reality scene (water maze)

It is worth to compare to the GSR signals for both patients in Fig. 5.5. The performance of two different probands in the same reference scene is unequal. Red markers (stress detection made by the therapist) are visible in both cases. One of the probands could finish the same task in shorter time than the other one. Many differences are visible, for example, different trend of GRS curve. Base line of GRS curve differs between individual probands even during quiet stance or relaxation. Therefore, interindividual comparison and data analysis of more probands and measurements in different conditions are necessary for optimal adjustment of the therapeutic process for each individual patient.

The idea central to the data approach presented in this chapter is to integrate time-synchronized data from many different sources and to analyze the data in order to better understand how the therapy or rehabilitation session affects the proband. The physiological data shown above represent only one possible facet of all data that may be relevant to the therapy session process. Other data sources may include:

- Low level data available from the input devices used throughout the training session—in case of the balance force platform, the weight load sensor data; in case of keyboard and mouse in more traditional settings, the keys pressed and the mouse positions in time, etc. This may seem redundant because the input data are further processed by the virtual scene engines, however, some observations may be done even on this low-level data, such as detecting of tremors.
- A record of the actions the proband took in terms of the virtual environment used (i.e. the chessboard scene or the 3D virtual environment)—which (sub)goals the proband completed in which time, the possible errors made by the proband, etc. A more detailed example of this data will be given in the next section.
- A video recording the entire training session. An appropriate post-analysis of the video can discover significant movement events, evaluate facial expressions, discover specific emotions, and correlate it with the sensor data.
- A journal of events that arose during the training sessions, that may help to identify time intervals where the physio/game data may be corrupted or unreliable (either by sensors being reattached/adjusted by the proband, or by outside disturbances). This can be realized in the simplest form as a series of time marks specified by the tester/therapist; the current time interval can be marked to indicate that the data in that moment may be unreliable. This approach to event journaling was used in the training sessions using the balance force platform. More complicated approaches may employ a more complex hierarchy of events and associated annotations.

5.4 Data Processing and Visualization

Given the raw data that can be obtained from various sensors described previously, we will now focus on the description of data structures and data flows integral to our approach. This will be demonstrated on use-cases mentioned above—the chessboard scene and virtual reality scenes. One minor terminology remark: since many of the scenes considered here for training and rehabilitation have game-like character and using 2D/3D game engines for scene visualization is quite common, we will refer to the testing applications with virtual environment as to games (as in serious games [7] and use game-related terminology further on. Indeed, many of the scenarios used in training and rehabilitation applications resemble computer games: there are given goals that the proband should accomplish, the proband's progress in the training session is measured by some kind of score (based on completion time, number of errors the proband made, etc.). The scenes can have adjustable difficulty level (by setting time limits, making scenes more complicated to navigate, etc.). Therefore, using serious game approach to the training and rehabilitation applications seems very natural. Conversely, many training and rehabilitation approaches that employ software applications can be gamified, by turning them into repeatable scenarios all with difficulty levels, awards and scores.

5.4.1 Data Processing

Since the raw data acquired from sensors are quite low-level and use arbitrary device-related units, it is necessary to perform some preprocessing, followed by peak detection/feature extraction steps. In this section, the preprocessing of the sensory data originated from Homebalance platform and the structure of the game related data for games used in the scenarios mentioned before as well as higher-level data that may be derived from the previous two sources will be discussed.

5.4.1.1 ECG

Electrocardiogram (ECG) represents electrical activity of human heart, typically measured by electrodes from the human body. ECG is composite from 5 waves—P, Q, R, S and T. ECG signal, after necessary denoising, can be used to extract many useful health indicators, most notably the heart rate frequency. Many algorithms for heart rate detection are based on QRS complex detection and the heart rate is estimated based on the distance between QRS complexes. The example of heart rate signal detected from source ECG signal can be seen in Fig. 5.6.

There are many algorithms for QRS complex detection, perhaps the best known is the on-line real-time algorithm by Pan and Tompkins [21].

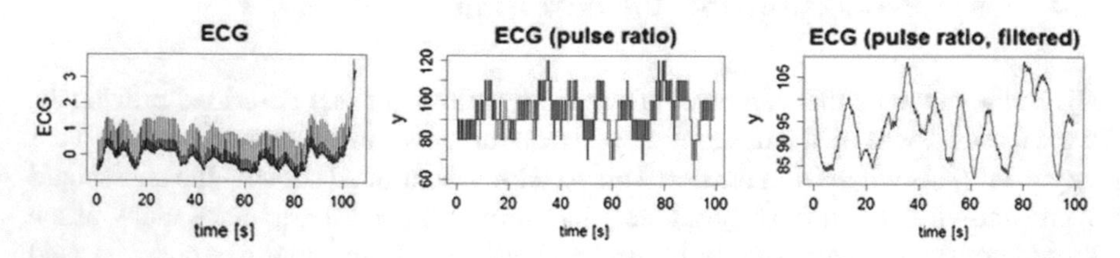

Fig. 5.6 ECG signal and the estimated heart rate frequency by using a sliding window algorithm for counting QRS complexes

5.4.1.2 EIMP

Measurement of the electrical impedance (EIMP) is a method in which the electrical conductivity, permittivity, and impedance of a part of the body is inferred from surface electrode measurements. Lung resistivity increases and decreases several-fold between inspiration and expiration. The investigated parameter is breathing curve derived from changes in the bioimpedance of the chest. The benefit of this method is non-invasive measurement of basic physiological function. Increase of breathing frequency is a sign of high difficulty of the exercise.

5.4.1.3 GSR

The galvanic skin response (GSR), also known as electrodermal activity or skin conductance response, is defined as a change in the electrical properties of the skin. The signal can be used for capturing the autonomic nerve responses as a parameter of the sweat gland function. GSR can be useful for estimating stress levels and cognitive load [1]. GSR has been closely linked to autonomic emotional and cognitive processing, and can be used as a sensitive index of emotional processing and sympathetic activity.

GSR signal processing is by no means a trivial task. External factors such as temperature and humidity affect GSR measurements, which can lead to inconsistent results. Internal factors such as medications can also change GSR measurements, demonstrating inconsistency with the same stimulus level. Also, galvanic skin responses are delayed, typically 1–3 s. There is a tremendous variability across individuals in terms of galvanic response; many individuals exhibit extremely low galvanic response to cues, while others respond extremely strongly. This suggests that GSR would be useful mainly to compare stress levels in various stages of training/rehabilitation of a single individual.

5.4.1.4 Game Events

Aside from the data from sensors, it is important to pay attention to what exactly is going on in the virtual scene, in the given moment. Therefore, it is useful to record:

- Game event data to be able later to reconstruct the game state and to identify the moments where the player has to solve various problems (find a way over the street in virtual reality scene, stabilize a sphere over the target position for a specified duration in the chessboard scene) so they can be correlated with physiological measures (Is the moment particularly stressful? Did the player's heart rate rise?).
- Low-level input data, which can be used for detecting anomalies from the normal device usage (e.g. by keeping track of mouse position when using it as an input device it is possible to detect abnormalities that may be evidence of tremors, etc.).

- The exact structure of game event data (or alternatively, the game model) depends heavily on the scenario used, however, the game event data together with associated data flows and evaluation logic share a common structure for all scenarios, with three main components: the task assignment, the gameplay events and the game logic itself.

The task assignment is a collection of game configuration parameters. More generally, the task assignment describes the game goal or mission to accomplish together with various settings that affect the course of game scenario. In the chessboard scenario, the data describing the task assignment contain attributes with their default values as follows:

- board_size = Vector2d(8,8)—a chessboard has 8 rows and 8 columns
- starting_position = Vector2d(3,4)—initial position of the ball
- time_required_to_stay_in_the_goal = 1000 (ms)—time needed to activate the goal
- goal_positions = list(Vector2d(1,3), Vector2d(3,5)…)—list of pre-generated positions.

The gameplay events represent occurrences of various important points during gameplay. All events contain UTC timestamps, some events may contain additional data. The event flow is considered as a companion data source to the sensors mentioned above (or, as another virtual sensor) and the event data are later coupled and evaluated together with all physiological data gathered during the sessions. The record of event data (the event log) is a rich source of information about the course of gameplay, its difficulty and the steps the player chose to deal with the challenges presented. The events generated in the chessboard game scenario together with the data they carry are:

- game_started (timestamp)
- ball_moved_to (timestamp, goal_position) – triggered when ball changes position
- goal_reached (timestamp, goal_position, goal_index) – triggered when a goal is activated
- game_finished (timestamp)—triggered when the last goal is activated.

The third component necessary to fully describe the game is the game logic itself. The game algorithm for the chessboard scene, expressed in a very informal pseudocode, is shown below.

```
Chessboard_Scene_Pseudocode:

    create board with size = board_size
    create ball with position = starting_position
    trigger_event(game_started)
    while not empty(goal_positions):
        goal = remove_first(goal_positions)
        wait for:
            (ball_position == goal) and
            (stay_duration == time_required_to_stay_in_the_goal)
        trigger_event(goal_reached)
    trigger_event(game_finished)
```

The three components together comprise the game model and sufficiently describe the inner workings of the game, allowing to formalize the format of input (task assignment) data, output (game events, game outcomes) data, and the implicit or explicit relationships between the two (game logic).

This knowledge is sufficient for reconstructing and evaluating the exact progress of the gameplay later, perhaps for purposes of finding the moments where the player experienced major struggles in the game and adjusting the game difficulty for later plays accordingly in order to optimize the training/learning process that the game is designed to facilitate.

5.4.1.5 Higher-Level Data

From the low-level data mentioned above, various high-level indicators can be derived. GSR, for instance, is a good resource for estimating stress levels [26]. It is also a prospective physiological indicator of cognitive load, defined as the amount of information that the short-term (or working) memory can hold at one time. Cognitive Load Theory explained by Sweller and his colleagues in [27] suggests that learners can absorb and retain information effectively only if it is provided in such a way that it does not "overload" their mental capacity.

Monitoring of physiological functions is useful tool for early detection of increasing stress level of the patient during the therapeutic process. Monitoring of the patient leads to detection of change of the patient's state even if it is not visible for the therapist. Normal level of heart rate and other parameters is individual for each patient. It is useful to collect data before the start of the therapeutic intervention, during the relaxation phase. The patient should be in optimal stress level during the whole therapeutic process. The current stress level should be compared to normal level of the individual patient and to optimal stress level of similar patients from the database.

Sensor data are managed by computers (stored, processed, analyzed, etc.) as time-series data. The PMML (Predictive Markup Language) format [22] is a suitable form for representation of these signal data and other statistical and data mining models computed and stored in our dataspace.

5.4.2 Data Visualization

Properly visualized data offer very good insight into the problems the patient has to face when coping with a situation he/she considers to be difficult. It can help to distinguish between the situation when the patient is not able to ensure some specific movement and the situation when the problem is caused by his/her psychological block.

Figure 5.7 visualizes several features derived from the raw data discussed in the previous paragraphs: the pulse rate (computed from the raw ECG data), GSR and EIMP data measured during a chessboard scene. The vertical red lines indicate occurrence of unexpected events from outside the training scenario that may affect the proband's performance, the vertical blue lines indicate important time points in the testing scenario—in this case, the time when the proband successfully completes a sub-goal of the game scenario (in the chessboard scene, the moment where the sphere is moved to the target position). The bottom two graphs show the ideal XY position of the sphere as if the proband chose the straightest path to the next target position without any deviation. The subjective observation of the therapist (insertion of red lines) may differ from the internal status of the patient. Sometimes, the therapist can assume entering a stress event, but the sensors do not confirm it and vice versa. The sensor data could be correlated with other data to explain the stress situation and the reason for it. For example, the patient doesn't trust himself to perform an exercise involving standing on heels (only thinking about it would cause stress); but he/she is able to do it (sensors can help to discover it). Helpful would be using, e.g., a Kinect camera to monitor patient movements and evaluate facial expressions—see, e.g., paper [10].

5.5 Supporting Decisions with Extracted Knowledge

Observations on the data from the Figs. 5.5, 5.6 and 5.7 are the result of a long-term experience gained by a highly interested susceptive therapist who has learned from his/her patients about some interesting features he/she should look for in the data of new patients. But there are certainly also additional features that remain still unknown and that could point e.g. to some hidden patient's problems that have to be seriously considered before significant improvement in the therapy can be reached. The way towards finding new approaches to interpretation of patient´s data has to be based on careful analysis and evaluation of data obtained from a much bigger group of patients than that of a single therapist or even of a single hospital. This can be archived through the architecture designed in this article. In the following, we report on several therapy successes achieved by applying the discussed approach. In the 4-day case study discussed below, the patient with balance disorder after traumatic brain injury subjectively claims that he/she has bad stability at shifting/transferring weight to the heals. His/her training task is presented by the chessboard (Fig. 5.8).

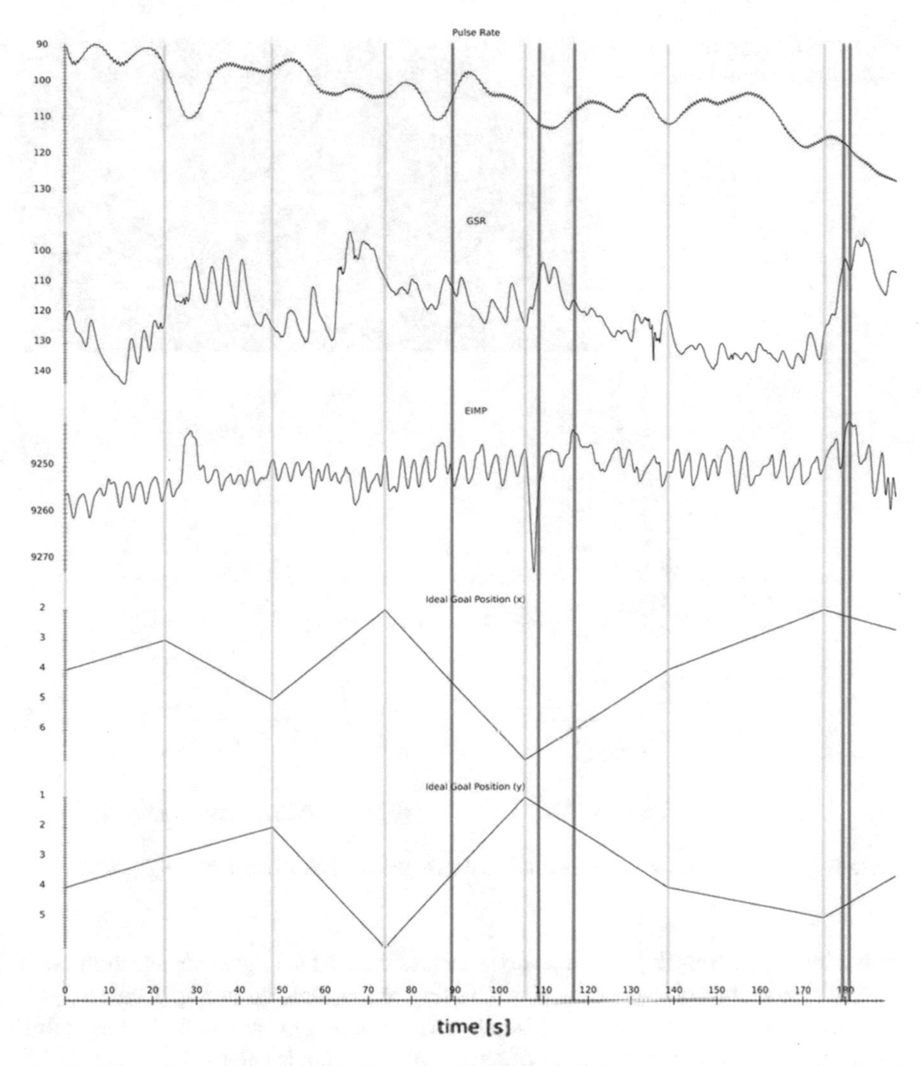

Fig. 5.7 Visualization of selected features after data preprocessing

The numbers, 1–6, denote board positions, the sequence of required movements, of the globe that correspond to the projections of shifting the center of gravidity. Observation details follow.

Day 1: A high sensitivity of the platform is set—this means that very small change of patient´s center of gravity position results in a move of the globe on the game screen. The patient is not able to transfer his/her weight using a smooth or straight movement – this is achieved in minor independent steps and consequently the trajectory of the projection of the gravity center is long and has zigzag form. He/she is not

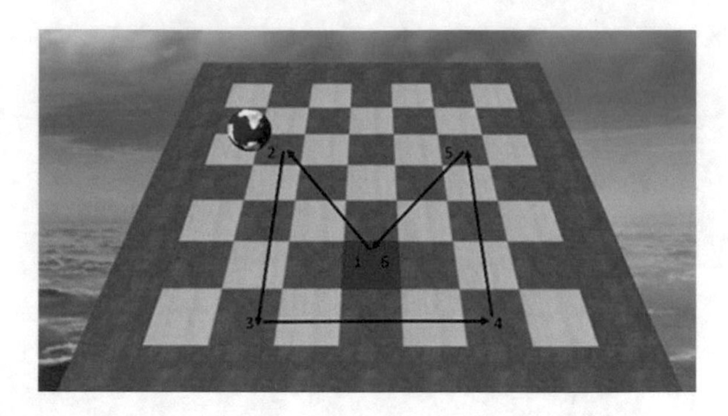

Fig. 5.8 Trajectory of the required globe movement

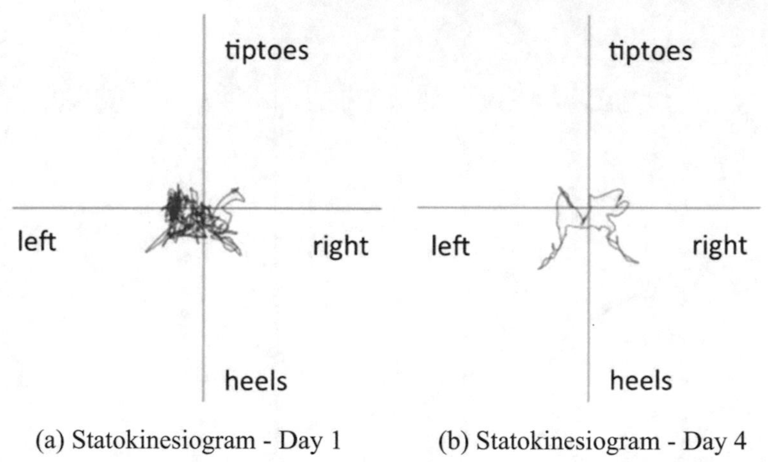

Fig. 5.9 Statokinesiograms that correspond to the chessboard task illustrated in Fig. 5.8

able to remain (persist in) on the position longer than 0.1 s. Higher stress parameters are indicated. The corresponding statokinesiogram is depicted in Fig. 5.9a.

During the following therapies, the platform sensitivity is gradually lowered; this means that the movement extent required from the patient is enlarged. Moreover, the time the patient is expected to remain (persist in) on any position is increased. This includes the posture on heels as well.

Day 4: The patient is already able to stand on heels for 3 s. The trajectory has changed significantly—it is much straighter and shorter compared to that of the first day. Even maximum movement extent is larger and stress parameters are lower. The corresponding statokinesiogram is depicted in Fig. 5.9b. This patient succeeded to achieve improvement in the heel standing very quickly compared to the other patients and simultaneously his/her stress level was reduced. These facts support an assumption that at the beginning, there was a high anxiety caused by patient's lack of confidence and the resulting stress influenced the patient´s performance, even though motoric deficit was not so big that he/she could not master the given task to the end. The therapist shares this observation with the patient and supports his claim

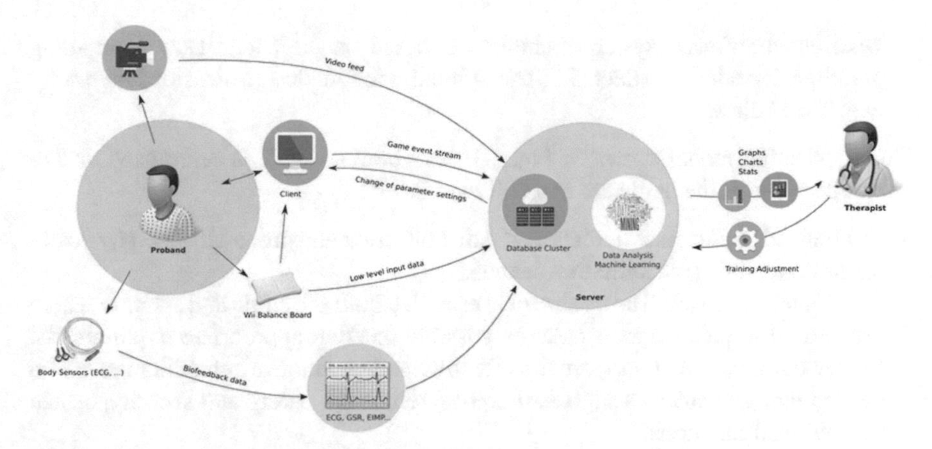

Fig. 5.10 Application model architecture

by calling the patient's attention to both kinesiograms on Fig. 5.9 that are carefully explained to the patient. This illustrative therapy is well understood by the patient who can clearly see his/her advance. This approach also increases motivation of the patient to continue in the rehabilitation process. In this way, visual feedback can significantly speed-up the improvement of the patient.

In another case study [28], a group of 16 patients with balance disorder after brain injury was studied. The group showed improvement, especially in dynamic stabilometric parameters. A survey has shown, in the most of patients, an improvement of their condition in connection with the stability disorder, which they reported before the treatment.

The mean improvement in Berg Balance Scale test [23] mentioned in Sect. 5.2 in the examined group of patients was 12.3%. Specifically, a patient from the study was evaluated with 47 points before therapeutic sessions and with 52 after them.

5.6 Overall System Architecture

After presenting the functionality that has been already experimentally used and tested by a set of selected patients, we now show how the realized functional modules are included in the overall architecture, which in this section is presented at two levels that also represent two logical design phases:

- Application model architecture—it gives an overview of participating human actors, data sources and data flows; it is depicted in Fig. 5.10, which is self-explanatory and visually supports the approach presentation we provided in the previous sections.

- Distributed software system architecture, based on the Cloud-Dew Computing paradigm introduced in Sect. 5.1. The rationales for our design decisions are briefly explained below.

The application model shown in Fig. 5.10 can be realized as a centralized or distributed system at the hardware and software level:

- Centralized approach. All treatment, data collection and processing are physically placed at a clinic (rehabilitation center).
- Distributed approach. The treatment is provided both at a clinic and at home. There are several implementation patterns possible meeting appropriate requirements. In our case, we need support for effective and productive rehabilitation and a system that guarantees a high availability, reliability, safety and security, beside its advanced functionality.

We have focused on Cloud-Dew computing to meet the above objectives. Our long-term goal is to have a network of cooperating nodes, where a node denotes a cloud server associated with a clinic (rehabilitation center) steering a set of home infrastructures equipped with balance force platforms and computing, storage and sensor devices. Each home system works autonomously, even if the Internet connection is not available now, and in specific intervals exchanges collaborative information (analysis results, data mining models, etc.) with the center involved in a cloud; here, appropriate security rules are preserved. The role of the processing on the premise is also annotation of the collected data with metadata before sending it to the cloud associated with the center.

The above cloud-dew concepts are reflected in Fig. 5.11. The Cloud Server coordinates activities of several home infrastructures, simply Homes. Client Program participates in the traditional software pattern Client-Server. Moreover, it includes the software functionality associated with on-premise data processing. Sensor Network Management steers the dataflow from sensors and other data stream sources (e.g. video cameras) to the Home Dataspace managed by the Data Space Management System (DSMS). The Dew Server acts as a specific proxy of the Cloud Server; among others, it is responsible for data synchronization on the Home side.

The Cloud Server may be connected to other cloud servers within a SKY infrastructure [16].

5.7 Conclusions

Rehabilitation of balance disorders, a significant category of brain damage impairments, is an issue of high social importance involving many research and development challenges. Patient rehabilitation and exercise supervised by an experienced therapist is an ultimate remedy to most of these balance disorders. This process often requires lot of time during which patient's improvement is very slow. Under such conditions, it is not easy to motivate the patient to adhere to the recommended exercise. We have

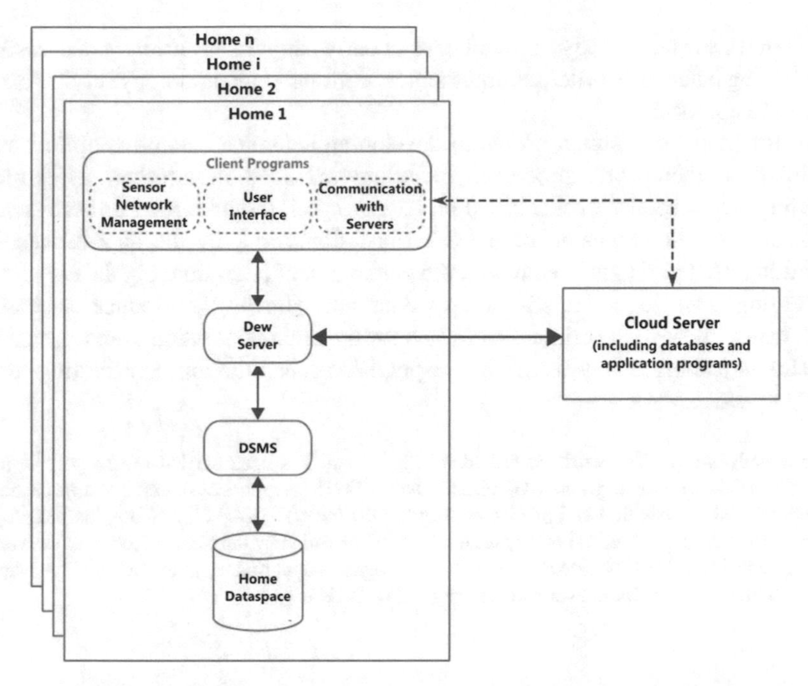

Fig. 5.11 Cloud-dew architecture, where dew is represented by the home infrastructure. Based on Wang [29, 30]

developed for that purpose a game like environment in which the patient uses the WiiFit-based balance force platform to control computer games carefully designed to tempt the patient to repeat movements recommended by the therapist. An appropriately selected computer game offers to the patient continuous visual feedback that indicates how far he/she is from the target location. There are mentioned briefly some case studies that prove advantages of this approach and indicate that success of this type of therapy relies on highly individualized approach to the patient that can currently be ensured by an experienced therapist, only. This cannot be provided on a mass scale, of course. In this chapter, we have introduced a promising Big-data centric therapy approach that can take over some duties of the experienced therapist without compromising individual needs of the patient.

Our key objectives included: (a) Providing the patient with advanced training means-this was achieved by the WiiFit-based balance force platform with wireless connection to a computing platform equipped by intelligent software; (b) Enabling unlimited access to this training system-a home version of the system is provided to patients in the hospital; (c) Increasing the effect and productivity of the therapy-this is achieved by capturing a set of physiological parameters by a network of sensors, processing and analyzing this data and consequent optimizing therapy parameters setting; and (d) Final technical realization of the proposed functionality by means that guarantee a high reliability, safety and security – this is provided by our focus on the Cloud-Dew technology. A small core of early adopters is currently successfully conducting balance disorder rehabilitation according to methodology relying on the proposed approach.

In the future research, we plan to develop an automated adaptive optimization of involved therapeutic processes including extended monitoring, assessment, learning, and response cycles based on patient-specific modeling. Further research plans include the extension of the described framework by the data provenance functionality. This is associated with the generation of additional big data resources increasing trust to collected therapy data and allowing reproduce successful treatments, improve the clinical pathways for the brain restoration domain, etc.

This is pathing a way to the future precision rehabilitation contributing to the greater good.

Acknowledgements The work described in this chapter has been carried out as part of three projects, namely, research grants SGS16/231/OHK3/3T/13 "Support of interactive approaches to biomedical data acquisition and processing" and SGS17/206/OHK4/3T/17 "Complex monitoring of the patient during the virtual reality based therapy" provided by the Czech Technical University in Prague and the Czech National Sustainability Program supported by grant LO1401 "Advanced Wireless Technologies for Clever Engineering (ADWICE)".

References

1. J. Bakker, M. Pechenizkiy, N. Sidorova, What's your current stress level? Detection of stress patterns from GSR sensor data, in *Data Mining Workshops IEEE 11th International Conference* (2011), pp. 573–580
2. A. Bohuncak, M. Ticha, M. Janatova, Comparative study of two stabilometric platforms for the application in 3D biofeedback system in *Abstracts of the 6th international posture symposium*, p. 21
3. A. Bohuncak, M. Janatova, M. Ticha, O. Svestkova, K. Hana, Development of interactive rehabilitation devices, in *Smart Homes* (2012), pp. 29–31
4. N.A. Borghese, M. Pirovano, P.L. Lanzi, S. Wüest, E.D. de Bruin, Computational intelligence and game design for effective at-home stroke rehabilitation. Games Health: Res. Dev. Clin. Appl. **2**(2), 81–88 (2013)
5. O. Cakrt et al., Balance rehabilitation therapy by tongue electrotactile biofeedback in patients with degenerative cerebellar disease. NeuroRehabilitation **31**(4), 429–434 (2012)
6. K.H. Cho, K.J. Lee, C.H. Song, Virtual-reality balance training with a video-game system improves dynamic balance in chronic stroke patients. Tohoku J. Exp. Med. **228**(1), 69–74 (2012)
7. R. Dörner, S. Göbel, *Serious Games: Foundations: Concepts and Practice* (Springer, Cham, 2016), p. 2016
8. I. Elsayed, Dataspace support platform for e-science. Ph.D. thesis, Faculty of Computer Science, University of Vienna, 2011. Supervised by P. Brezany, Revised version published by Südwestdeutscher Verlag für Hochschulschriften (https://www.svh-verlag.de/), 2013. ISBN: 978-3838131573, 2013
9. J.-F. Esculier et al., Home-based balance training programme using WiiFit with balance board for Parkinson's disease: a pilot study. J. Rehabil. Med. **44**, 144–150 (2012)
10. M. Ferreira, A. Carreiro, A. Damasceno, Gesture analysis algorithms. Procedia Technol. **9**, 1273–1281 (2013)

11. Force Platform (2016), https://en.wikipedia.org/wiki/Force_platform. Accessed 12 Nov 2016
12. V. Gatica-Rojas, G. Méndez-Rebolledo, Virtual reality interface devices in the reorganization of neural networks in the brain of patients with neurological diseases. Neural Regeneration Res. **9**(8), 888–896 (2014)
13. O.M. Giggins, U.M. Persson, B. Caulfield, Biofeedback in rehabilitation. J. Neuroeng. Rehabil. **10**(1), 60 (2013)
14. J.A. Gil-Gómez, R. Lloréns, M. Alcañiz, C. Colomer, Effectiveness of a Wii balance board-based system (eBaViR) for balance rehabilitation: a pilot randomized clinical trial in patients with acquired brain injury. J. Neuroeng. Rehabil. **8**(1), 30 (2011)
15. M. Janatová, M. Tichá, M. Gerlichová et al., Terapie poruch rovnováhy u pacientky po cévní mozkové příhodě s využitím vizuální zpětné vazby a stabilometrické plošiny v domácím prostředí. Rehabilitácia **52**(3), 140–146 (2015)
16. K. Keahey, M. Tsugawa, A. Matsunaga, J. Fortes, Sky computing. IEEE Internet Comput. **13**(2009), 43–51 (2009)
17. B.B. Lahiri, S. Bagavathiappan, T. Jayakumar, J. Philip, Medical applications of infrared thermography: a review. Infrared Phys. Technol. **55**(4), 221–235 (2012)
18. V. Mayer-Schonberger, K. Cukier, *Big Data: A Revolution That Will Transform How We Live, Work and Think* (John Murray (Publishers), London, 2013)
19. T. O'Donovan, J. O'Donoghue, C. Sreenan, P. O'Reilly, D. Sammon, K. O'Connor, A context aware wireless body area network (BAN), in *Proceedings of the Pervasive Health Conference* (2009)
20. M. Oliver, et al., in *Smart Computer-Assisted Cognitive Rehabilitation for the Ageing Population, Ambient Intelligence-Software and Applications—7th International Symposium on Ambient Intelligence*, vol. 476 of the Series Advances in Intelligent Systems and Computing (2016), pp. 197–205
21. J. Pan, J.W. Tompkins, A real-time QRS detection algorithm. IEEE Trans. Biomed. Eng. BME **32**(3), 230–236 (1985)
22. PMML, (2016), http://dmg.org/pmml/v4-3/TimeSeriesModel.html. Accessed 25 Feb 2017
23. Rehab-Measures, (2017), http://www.rehabmeasures.org/Lists/RehabMeasures/PrintView.aspx?ID=888. Accessed 27 Feb 2017
24. K. Skala, D. Davidovic, E. Afgan, I. Sovic, Z. Sojat, Scalable distributed computing hierarchy: cloud, fog and dew computing. Open J. Cloud Comput. (OJCC), **2**(1), 16–24 (2015)
25. Statokinesiogram-Definition, (2017). https://www.omicsonline.org/open-access/.comparative-study-using-functional-and-stabilometric-evaluation-ofbalance-in-elderly-submitted-to-conventional-physiotherapy-and-w-jppr-1000109.pdf. Accessed 27 Feb 2017
26. F. Sun, C. Kuo, H. Cheng, S. Buthpitiya, P. Collins, M. Griss, in *Activity-Aware Mental Stress Detection Using Physiological Sensors*. Lecture Notes of the Institute for Computer Sciences. Social Informatics and Telecommunications Engineering Mobile Computing, Applications, and Services (2012), pp. 211–230
27. J. Sweller, P. Ayres, S. Kalyuga: *Cognitive Load Theory*, Springer Science & Business Media, (2011)
28. M. Tichá, M. Janatová, R. Kliment, O. Švestková, K. Hána, Mobile rehabilitation device for balance training with visual feedback, in *Proceedings of International Conference on Mobile and Information Technologies in Medicine and Health* (2014), pp. 22–24
29. Y. Wang, Cloud-dew architecture. Int. J. Cloud Comput. **4**(3), 199–210 (2015a)
30. Y. Wang, The initial definition of dew computing. Dew Comput. Res. (2015b)
31. Y. Wang, Definition and categorization of dew computing. Open J. Cloud Comput. (OJCC), **3**(1), 1–7 (2016)

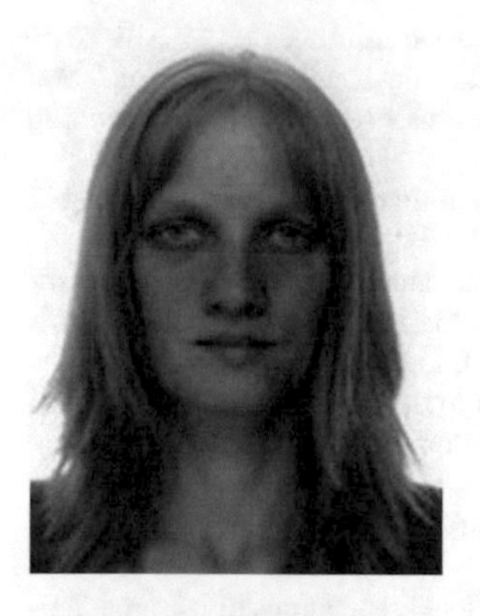

Dr. Marketa Janatova graduated from the 1st Faculty of Medicine at the Charles University in Prague in 2008. Since 2010 she works at the Department of Rehabilitation Medicine of the First Faculty of Medicine of Charles University and General Teaching Hospital in Prague as a medical doctor. Since 2011 she is a member of interdisciplinary team at Joint Department of Biomedical Engineering, Czech Technical University with a specialization in development of innovative technical devices for rehabilitation. Since 2013 she works at Spin-off company and research results commercialization center at the 1st Faculty of Medicine, Charles University in Prague. She participated in more than 30 publications and papers focused on development, validation and application of ICT solutions in diagnostics, telemonitoring and neurorehabilitation.

Miroslav Uller graduated from the Faculty of Electrical Engineering of the Czech Technical University in Prague in 2005. His interests include computer graphics, user interfaces, game programming and exotic programming languages. In 2010-2014 he participated in the European projects OLDES and SPES, aimed at developing creative software solutions to aim elderly people and people with disabilities. He currently works in the Department of Neurology in the Motol Hospital in Prague on processing of brain MRI data and at the Czech Institute of Informatics, Robotics and Cybernetics (CIIRC) of the Czech Technical University on software development for intelligent industrial applications.

Olga Stepankova graduated from the Faculty of Mathematics and Physics of Charles University and she successfully concluded her postgraduate studies in the field of mathematical logic in 1981. Since 1988 she has been working at the Faculty of Electrical Engineering of the Czech Technical University in Prague, where she became full professor in 1998. From 2016, she is the head of the department Biomedical Engineering and Assistive Technology of the Czech Institute of Intormatics, Robotics and Cybernetics (CIIRC). Her research is dedicated to data mining, artificial intelligence and development of assistive technologies.

Prof. Peter Brezany serves as a Professor of Computer Science at the University of Vienna, Austria and the Technical University of Brno, Czech Republic. He completed his doctorate in computer science at the Slovak Technical University Bratislava. His focus was first on the programming language and compiler design, automatic parallelization of sequential programs and later high-performance input/output, parallel and distributed data analytics, and other aspects of data-intensive computing. Since 2001 he has led national and international projects developing Grid- and Cloud-based infrastructures for high-productivity data analytics. Dr. Brezany has published one book, several book chapters and over 130 papers. He co-authored a book addressing big data challenges and advanced e-Science data analytics; it was published in Wiley in 2013.

Marek Lenart is an M.Sc. candidate at the University of Vienna, at the Faculty of Computer Science. He was involved in the European projects ADMIRE (Advanced Data Mining and Integration Research for Europe) and SPES (Support Patients through E-service Solutions) and co-authored several publications addressing the results of these projects. His focus was on advanced visualization of models produced by mining big scientific and business data. Now he works on ambitious industrial projects that are focused on simulation, analysis and optimization of data related to car production. The key objective of this effort is to minimize carbon dioxide emission in the context of the EU and international regulations.

Chapter 6
Big Data in Agricultural and Food Research: Challenges and Opportunities of an Integrated Big Data E-infrastructure

Pythagoras Karampiperis, Rob Lokers, Pascal Neveu, Odile Hologne, George Kakaletris, Leonardo Candela, Matthias Filter, Nikos Manouselis, Maritina Stavrakaki and Panagiotis Zervas

Abstract Agricultural and food research are increasingly becoming fields where data acquisition, processing, and analytics play a major role in the provision and application of novel methods in the general context of agri-food practices. The chapter focuses on the presentation of an innovative, holistic e-infrastructure solution that aims to enable researches for distinct but interconnected domains to share data, algorithms and results in a scalable and efficient fashion. It furthermore discusses on the potentially significant impact that such infrastructures can have on agriculture and food management and policy making, by applying the proposed solution in variegating agri-food related domains.

P. Karampiperis · N. Manouselis · M. Stavrakaki · P. Zervas (✉)
Agroknow, 110 Pentelis Street, 15126 Marousi, Greece
e-mail: pzervas@agroknow.com

R. Lokers
Wageningen Environmental Research - WUR, P.O. Box 47, 6700 AA Wageningen, The Netherlands

P. Neveu
Institut national de la recherche agronomique (INRA), 2 Place Pierre Viala, 34060 Montpellier Cedex 02,, France

O. Hologne
Institut national de la recherche agronomique (INRA), INRA RD 10, Route de Saint-Cyr, Versailles 78026, France

G. Kakaletris
Department of Informatics and Telecommunications, National and Kapodistrian University of Athens, Panepistimiopolis, Ilisia, 15784 Athens, Greece

L. Candela
Institute of the National Research Council of Italy, Area della Ricerca CNR, via G. Moruzzi 1, Pisa 56124, Italy

M. Filter
Federal Institute for Risk Assessment, Max-Dohrn-Str. 8-10D, 10589 Berlin, Germany

A. Emrouznejad and V. Charles (eds.), *Big Data for the Greater Good*, Studies in Big Data 42, https://doi.org/10.1007/978-3-319-93061-9_6

Keywords E-infrastructure · Agro-climatic · Economic modelling · Food security · Plant phenotyping · Food safety · Risk assessment

6.1 Introduction

Over the past years, big data in agricultural and food research has received increased attention. Farming has been empirically driven for over a century but the data collected was not digital. Agriculture Canada's family of research centres (circa 1920s) meticulously accounted for wheat yields across farms and weather patterns in order to increase efficiency in production. Big Data is different from this historic information gathering in terms of the volume and the analytical potential embedded in contemporary digital technologies. Big Data proponents promise a level of precision, information storage, processing and analysing that was previously impossible due to technological limitations [4].

Farmers have access to many data-intensive technologies to help them monitor and control weeds and pests, for example. In this sense, data collection, data modelling and analysis, and data sharing have become core challenges in weed control and crop protection [11]. As smart machines and sensors crop up on farms and farm data grow in quantity and scope, farming processes will become increasingly data-driven and data-enabled [13]. This is due to the fact that vast amounts of data are produced by digital and connected objects such as farm equipment, sensors in the fields or biochips in animals. Moreover, robots are becoming more and more popular, as it is well illustrated in dairy production. Alongside with the continuous monitoring that is producing well-structured data, other sources of data are produced and used. Interconnections of information systems and interoperability among them are also increasingly important. This is leading to new management modes of agricultural and food production, new services offered and new relationships along the supply chains regarding data sharing and re-use.

Current developments in ICT and big data science potentially provide innovative and more effective ways to support agricultural and food researchers to work with extremely large data sets and handle use cases involving big data. This new paradigm raises new research questions, new methods and approaches to perform research in agriculture and food. To support this paradigm, research facilities and e-infrastructures need to be revisited and new partnerships among academic institutions and private companies should be emerged. This will possibly lead to the involvement of scientists from areas of science and technology, who are not involved to the agricultural and food sector, to boost the innovation process in this sector. This also creates new challenges and opportunities for informed decision making from micro scale (i.e. precision agriculture) to macro scale (i.e. policy making).

Within this context, the main aim of this book chapter is twofold: (a) to present the design of AGINFRA+ , an integrated Big Data e-Infrastructure for supporting Agricultural and Food Research, which will be based on existing generic e-Infrastructures such as OpenAIRE (for publication and data set aggregation, indexing, mining and disambiguation), EUDAT (for cloud-hosted preservation and storage), EGI.eu (for cloud and grid resources for intensive computational applications), and D4Science (for data analytics); (b) to present three use cases for performing research in agriculture and food with the use of the proposed Big Data e-Infrastructure. More specifically, the first use case will be related to global agricultural modelling, intercomparison, and improvement of the research community that studies short and long-term food production under environmental and climate change conditions. The second use case will be related to addressing specific problems on food security, namely the need to efficiently analyze data produced by plant phenotyping and its correlation with crop yield, resource usage and local climates. The third use case will be related to the design of new high-performance food safety data processing workflows facilitating (i) the efficient extraction of data and models from the rich corpus of scientific literature, and (ii) the issue of generating easy-to-maintain open food safety model repositories.

This book chapter begins by presenting the suggested integrated big data infrastructure for agricultural and food research. It will then go on to provide an overview of the existing initiatives and e-infrastructures as well as details of the above-mentioned use cases. Finally, it will analyze the new challenges and directions that will arise for agriculture and food management and policing.

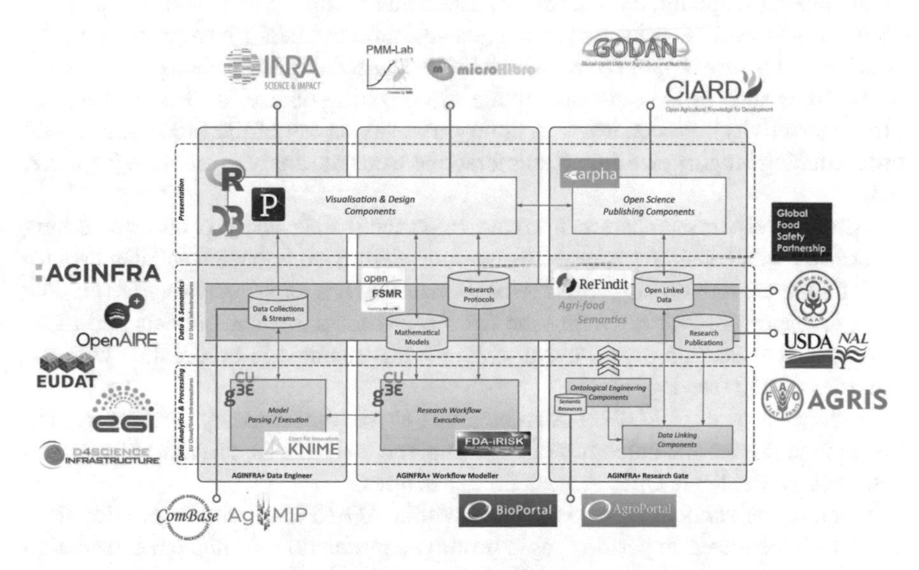

Fig. 6.1 Conceptual architecture showcasing how different technologies/tools form the vision of AGINFRA+

6.2 An Integrated Big Data Infrastructure for Agricultural and Food Research

AGINFRA+ aims to exploit core e-infrastructures such as EGI.eu, OpenAIRE, EUDAT and D4Science, towards the evolution of the AGINFRA data infrastructure, so as to provide a sustainable channel addressing adjacent but not fully connected user communities around Agriculture and Food.

To this end, AGINFRA+ entails the development and provision of the necessary specifications and components for allowing the rapid and intuitive development of variegating data analysis workflows, where the functionalities for data storage and indexing, algorithm execution, results visualization and deployment are provided by specialized services utilizing cloud based infrastructure(s). Furthermore, AGINFRA+ aspires to establish a framework facilitating the transparent documentation and exploitation and publication of research assets (datasets, mathematical models, software components results and publications) within AGINFRA, in order to enable their reuse and repurposing from the wider research community.

In a high-level, conceptual view on the AGINFRA architecture (see Fig. 6.1), the AGINFRA+ *Data and Semantics Layer* comprises the functionalities that facilitate the discovery and consumption of data collections (annotation, description enrichment, linking/mapping, etc.). These datasets and components enable researchers, innovators as well as businesses to compose customized data-driven services via the AGINFRA+ *Presentation Layer*, which builds upon the use of existing open source frameworks such as D3 (an open-source visualization library for data-driven real-time interactivity), Processing.js (a library that sits on top of the Processing visual programming language), Shiny (for interactive web data analysis) etc (see Figs. 6.2, 6.3).

Alternatively, researchers can create (advanced) data analysis and processing workflows which can be packaged as ready-to-deploy components and can be submitted for execution to the AGINFRA+ *Data Analytics and Processing Layer*. The layer is responsible for activating the necessary executable components and monitoring their execution over the available e-infrastructures in accordance with the provided design (see Fig. 6.4).

Finally, the evolved AGINFRA architecture entails the necessary tools for making the methodologies and outcomes (algorithms, research results, services) of research activities available via Open Access Infrastructures.

Overall, the services to be developed within AGINFRA+ aim to provide their users with the means to perform rapid prototype production, facilitate the execution of resource-intensive experiments, allow the agile and intuitive parameterization and repetition of experiments, and—ultimately—maximize the visibility and reuse of their research outcomes.

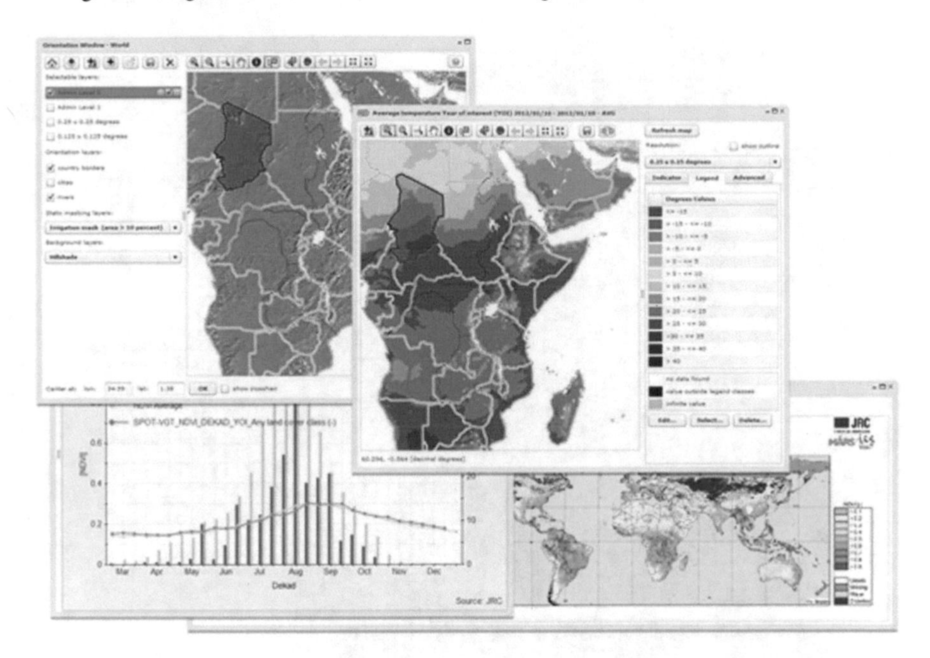

Fig. 6.2 Examples of various data types and sources used to support impact assessment

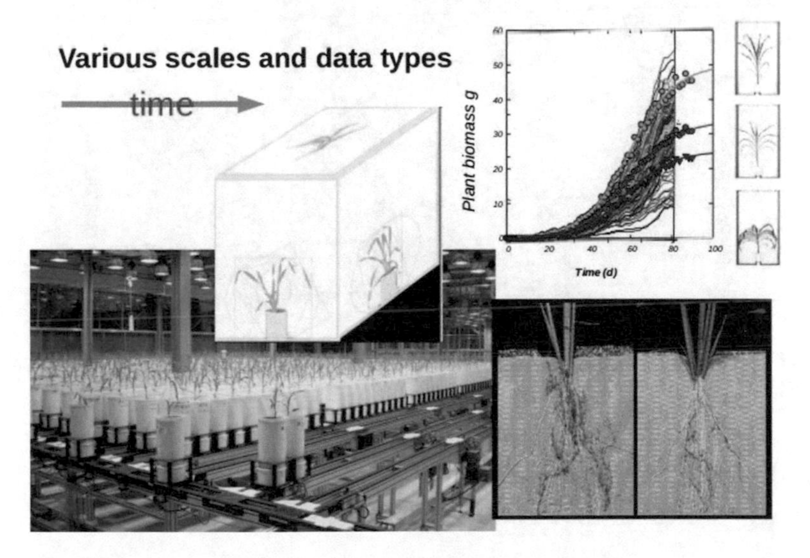

Fig. 6.3 Examples of visualization for high throughput phenotyping

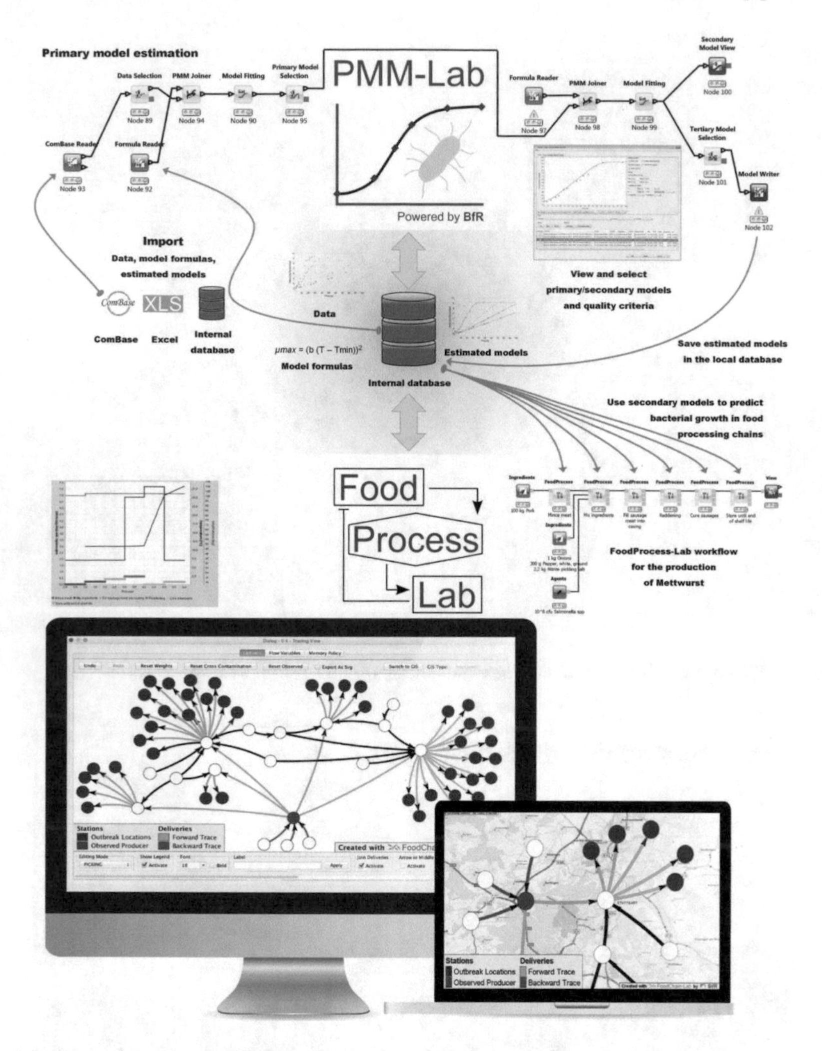

Fig. 6.4 Examples of graphical workflow based data analysis and visualisation features in FoodRisk-Labs

6.3 Related Work: Existing E-Infrastructures for Agricultural and Food Research

In this section, we provide an overview of existing initiatives and e-infrastructures that pertain to the vision and objectives of AGINFRA+.

6.3.1 AGINFRA

AGINFRA is the European research hub and thematic aggregator that catalogues and makes discoverable publications, data sets and software services developed by Horizon 2020 research projects on topics related to agriculture, food and the environment so that they are included in the European research e-infrastructure "European Open Science Cloud".

6.3.2 EGI-Engage

The EGI-Engage: Engaging the Research Community towards an Open Science Commons (https://www.egi.eu/about/egi-engage/) started in March 2015, as a collaborative effort involving more than 70 institutions in over 30 countries, coordinated by the European Grid Initiative (EGI) association. EGI-Engage aims to accelerate the implementation of the Open Science Commons by expanding the capabilities of a European backbone of federated services for compute, storage, data, communication, knowledge and expertise, complementing community-specific capabilities.

Agriculture, food and marine sciences are included as use cases in EGI-Engage, providing requirements that shape the new Open Science Commons platform. AGINFRA is positioned as the thematic data infrastructure and is exploring use cases and workflows of storing, preserving and processing data in the generic e-infrastructure.

6.3.3 OpenAire 2020

OpenAIRE2020 continues and extends OpenAIRE's scholarly communication open access infrastructure to manage and monitor the outcomes of EC-funded research. It combines its substantial networking capacities and technical capabilities to deliver a robust infrastructure offering support for the Open Access policies in Horizon 2020.

The integration of all AGINFRA scientific repositories in OpenAIRE is currently undergoing, aiming to have AGINFRA serving as the OpenAIRE thematic node for agriculture and food.

6.3.4 Eudat

EUDAT's vision is to enable European researchers and practitioners from any research discipline to preserve, find, access, and process data in a trusted environment, as part of a Collaborative Data Infrastructure (CDI) conceived as a network of collaborating, cooperating centres, combining the richness of numerous community-specific data repositories with the permanence and persistence of some of Europe's largest scientific data centres. EUDAT offers common data services, supporting multiple research communities as well as individuals, through a geographically distributed, resilient network of 35 European organisations.

AGINFRA+ partners have been early and continuously participating in EUDAT user workshops, providing requirements and use cases on how various EUDAT services may be used by our scientific communities.

6.3.5 D4Science

D4Science is a Hybrid Data Infrastructure servicing a number of Virtual Research Environments. Its development started in the context of the homonymous project cofounded by the European Commission and has been sustained and expanded by a number of EU-funded projects (ENVRI, EUBrazilOpenBio, iMarine). Currently, it serves as the backbone infrastructure for the BlueBRIDGE and ENVRIPlus projects.

AGINFRA+ partners are technology leaders in D4Science, while also contributing use cases and requirements to D4Science, through their participation of the corresponding projects or project events.

6.3.6 BlueBRIDGE

BlueBRIDGE aims at, innovating current practices in producing & delivering scientific knowledge advice to competent authorities and enlarges the spectrum of growth opportunities in distinctive Blue Growth areas and to further developing and exploiting the iMarine e-Infrastructure data services for an ecosystem approach to fisheries.

6.3.7 ENVRIPlus

ENVRIPLUS is a cluster of research infrastructures (RIs) for Environmental and Earth System sciences, built around ESFRI roadmap and associating leading e-infrastructures and Integrating Activities together with technical specialist partners.

Agriculture, food and marine sciences are represented as communities in ENVRIPlus

6.3.8 *Elixir*

The goal of ELIXIR is to orchestrate the collection, quality control and archiving of large amounts of biological data produced by life science experiments. Some of these datasets are highly specialised and would previously only have been available to researchers within the country in which they were generated.

For the first time, ELIXIR is creating an infrastructure—a kind of highway system—that integrates research data from all corners of Europe and ensures a seamless service provision that is easily accessible to all. In this way, open access to these rapidly expanding and critical datasets will facilitate discoveries that benefit humankind.

In the currently running EXCELERATE project (https://www.elixir-europe.org/excelerate) that develops further the ELIXIR infrastructure, a major use case is on Integrating Genomic and Phenotypic Data for Crop and Forest Plants.

6.3.9 *EOSC*

As part of the European Digital Single Market strategy, the European Open Science Cloud (EOSC) will raise research to the next level. It promotes not only scientific excellence and data reuse but also job growth and increased competitiveness in Europe, and drives Europe-wide cost efficiencies in scientific infrastructure through the promotion of interoperability on an unprecedented scale. It aims to bring together and align the existing core e-infrastructure services in order to smoothly support and integrate thematic and domain specific data infrastructures and VREs.

AGINFRA+ partners have already initiated a strategic discussion with the key players trying to shape the EOSC and have participated in the various workshops and brainstorming meetings.

6.4 Use Cases for Agricultural and Food Research

6.4.1 *Agro-Climatic and Economic Modelling*

6.4.1.1 Landscape

The case of a global agricultural modeling, intercomparison, and improvement research community that studies short and long-term food production under environmental and climate change conditions. In this case, the problem addressed is related to accessing, combining, processing and storing high volume, heterogeneous data related to agriculture/food production projections under different climate change scenarios, so that it becomes possible to assess food security, food safety and climate change impacts in an integrated manner, by a diverse research community of agricultural, climate and economic scientists.

The mission of this research community lies in improving historical analysis and short and long-term forecasts of agricultural production and its effects on food production and economy under dynamic and multi-variable climate change conditions, aggregating extremely large and heterogeneous observations and dynamic streams of agricultural, economical, ecophysiological, and weather data.

Bringing together researchers working on these problems from various perspectives (crop production and farm management methods, climate change monitoring, economic production models, food safety models), and accelerate user-driven innovation is a major challenge. The AGINFRA+ services will enable executable workflows for ecophysiological model intercomparisons driven by historical climate conditions using site-specific data on soils, management, socioeconomic drivers, and crop responses to climate. These intercomparisons are the basis for the future climate impact and adaptation scenarios: instead of relying on single model outputs, model-based uncertainties will be quantified using multi-model ensembles. The close interactions and the linkages between disciplinary models and scenarios, including climate, ecophysiology and socio-economics will allow researchers to prioritize necessary model improvements across the model spectrum.

Multi-model, multi-crop and multi-location simulation ensembles will be linked to multi-climate scenarios to perform consistent simulations of future climate change effects on local, regional, national, and global food production, food security, and poverty.

The data sources that can feed such bases required for this work are developed by different community members, are processed using different systems, and are shared among the community members. This creates several challenges that are connected to multiple factors: different platforms, diverse data management activities, distributed data processing and storage, heterogeneous data exchange, etc. and distributed model runs, data storage, scenario analysis, and visualization activities that take place. Thus, AGINFRA+ will also develop a reactive intensive data analysis layer over existing federations that will help the discovery, reuse and exploitation of heterogeneous data sources created in isolation, in very different and unforeseen ways in the rest of the communities' systems.

6.4.1.2 AGINFRA+ Advancement

Improving historical analysis and short and long-term forecasts of agricultural production and its effects on food production and economy under dynamic and multi-variable climate change conditions, is associated with several challenges in the area of Big Data. For short term (e.g. seasonal) forecasting of production timeliness is essential. To provide daily updates of forecasts, taking in the most actual information, real-time processing of up-to-date weather data and remote sensing data is required. For long-term projections, using among others climate ensembles from global or regional climate models, handling the volume of input and output data (combining, processing, storing) is more relevant, together with the availability of interconnected high-performance computing and cloud storage solutions.

As for now, many modelling exercises are performed "locally", mainly because the integrated e-infrastructure (meaning seamlessly interconnecting data-infrastructures, grid computing infrastructures and cloud-based infrastructures) are not available to the modelling community. Organizing such sustainable large-scale integrations and establishing working real-world modelling cases using these combined e-infrastructures would be a great leap forward for the involved research communities.

6.4.2 *Food Security*

6.4.2.1 Landscape

The AGINFRA+ infrastructure will be used to alleviate the big data challenges pertaining to specific problems on food security, namely the need to efficiently analyse data produced by plant phenotyping and its correlation with crop yield, resource usage, local climates etc. It will particularly explore how high throughput phenotyping can be supported in order to:

(a) Determine adaptation and tolerance to climate changes, a high priority is to design high-yielding varieties adapted to contrasting environmental conditions including those related to climate change and new agricultural management. It requires identifying of allelic variants with favourable traits amongst the thousands of varieties and natural accessions existing in genebanks. Genotyping (i.e. densely characterizing the genome of breeding lines with markers) has been industrialized and can now be performed at affordable cost and be able to link and analyse phenotyping and genotyping data is strategic for agriculture.
(b) Optimize use of natural resources. High throughput plant phenotyping aims to study plant growth and development from the dynamic interactions between the genetic background and the environment which plants develop (soil, water, climate, etc.). These interactions determine plant performance and productivity can be use in order to optimize and preserve natural resources.
(c) Maximize crop performance, gathering and analysing data from high throughput plant phenotyping allows a better knowledge of plants and their behaviour in specific resource conditions such as soil conditions and new climates.
(d) The tasks generally require intensive big data analysis as they present all the challenges associated with big data (Volume, Velocity, Variety, Validity).
(e) Each high throughput plant phenotyping produced several Tbytes of very heterogeneous data (sensor monitoring, images, spectrums). Data are produced at high frequencies and hundreds of thousands of images can be gather and be analysed each day. Such volumes require automatic data validation tools. On of major challenges of plant phenotyping is the Semantic Interoperability.

This AGINFRA+ use case will assess the effectiveness of the proposed framework in data intensive experimental sessions, where the distinct processing steps operate over different datasets, require synchronization of results from various partial computations, and use very large and/or streaming raw or processed data.

6.4.2.2 AGINFRA+ Advancement

Plant derived products are at the centre of grand challenges posed by increasing requirements for food, feed and raw materials. Integrating approaches across all scales from molecular to field applications are necessary to develop sustainable plant production with higher yield and using limited resources. While significant progress has been made in molecular and genetic approaches in recent years, the quantitative analysis of plant phenotypes—structure and function of plant—has become the major bottleneck.

Plant phenotyping is an emerging science that links genomics with plant ecophysiology and agronomy. The functional plant body (PHENOTYPE) is formed during plant growth and development from the dynamic interaction between the genetic background (GENOTYPE) and the physical world in which plants develop (ENVIRONMENT). These interactions determine plant performance and productivity measured as accumulated biomass and commercial yield and resource use efficiency.

Improving plant productivity is key to address major economic, ecological and societal challenges. A limited number of crops provides the resource for food and feed; reliable estimates indicate that food supplies need to be increased by quantity (50% by 2050) and quality to meet the increasing nutritional demand of the growing human (and animal) population. At the same time, plants are increasingly utilized as renewable energy source and as raw material for a new a generation of products. Climate change and scarcity of arable land constitute additional challenges for future scenarios of sustainable agricultural production. It is necessary and urgent to increase the impact of plant sciences through practical breeding for varieties with improved performance in agricultural environments.

The understanding of the link between genotype and phenotype is currently hampered by insufficient capacity (both technical and conceptual) of the plant science community to analyze the existing genetic resources for their interaction with the environment. Advances in plant phenotyping are therefore a key factor for success in modern breeding and basic plant research.

To deploy a "big data" strategy for organizing, annotating and storing phenotyping data for plant science in such a way that any scientist in Europe can potentially access the data of several for a common analysis. Which data should be standardized, which not? Standardization with common ontologies is crucial. Elaborate a consensus in the phenotyping community in Europe with high degree of standardization for environmental variables (types of sensors, units) and basic phenotypic data (dates, phenology) and higher flexibility for elaborate traits that need to reach maturity before being standardized. The relation with generalist databases (plant/crop ontologies) need to be established.

How to manage and process increasing data volume with distributed and reproducible data workflows? How to share data/how to access data from various experiments? How to combine large and heterogeneous datasets in an open data perspective? The most recent methods of information technologies will be used for smart data exchanges namely key discovery, matching and alignment methods for accessing and visualizing complex data sets in an integrative way.

6.4.3 Food Safety Risk Assessment

6.4.3.1 Landscape

In the context of the Food Safety Risk Assessment use case, the AGINFRA+ project will assess the usefulness of AGINFRA+ components and APIs to support data-intensive applications powered by the FoodRisk-Labs suite of software tools (https://foodrisklabs.bfr.bund.de). This includes the extension of FoodRisk-Labs' capabilities to handle large-scale datasets, to visualize complex data, mathematical models as well as simulation results and to deploy generated data processing workflows as web-based services.

More specifically, FoodRisk-Labs will be extended such that it can use and access AGINFRA Services. This will allow to design new high-performance food safety data processing workflows facilitating e.g. the efficient extraction of data and models from the rich corpus of scientific literature. Another workflow will address the issue of generation of easy-to-maintain open food safety model repositories (openFSMR), which exploit AGINFRA Ontological Engineering and Open Science Publishing Components. Mathematical models published in community driven openFSMR will then be used for large-scale quantitative microbial risk assessment (QMRA) simulations. These simulations incorporate predictive microbial models, models on food processing and transportation, dose response models as well as consumer behaviour models. AGINFRA components supporting the execution of computational intensive simulations as well as those helping to present simulation results will be applied here. Finally, preconfigured QMRA models will be deployed as easy-to-use web services using specialised AGINFRA components.

6.4.3.2 AGINFRA+ Advancement

The exploitation of food safety and food quality data using data-mining algorithms is a field of increasing relevance to food safety professionals and public authorities. Nowadays, the amount of experimental and analytical food safety data increase as well as information collected from sensors monitoring food production and transportation processes. However, there is a gap in free, easy-to-adapt software solutions that enables food safety professionals and authorities to exploit these data in a consistent manner applying state-of-the art data mining and modelling technologies. In addition, integrated, standardized data and knowledge management is required to establish quality-controlled model repositories that can be used for risk assessment and decision support.

The FoodRisk-Labs collection of software tools has been outlined right from the beginning as a community resource to allow broad application and joint developments. The specific food safety analysis and modelling functionalities were implemented as extensions to the professional open-source data analysis and machine learning platform KNIME (www.knime.org). The KNIME visual workflow composition user interface enables users to apply or adapt preconfigured data analysis workflows or to create new ones from the large number of available data processing modules ("nodes"). The selection of KNIME as the technical implementation framework guarantees important features like modularity, flexibility, scalability and extensibility. All FoodRisk-Labs KNIME extensions also provide preconfigured (empty) databases allowing users to easily manage their domain-specific data and/or establish model-based knowledge repositories.

6.5 New Challenges and Directions for Agriculture and Food Management and Policing

The variety of stakeholders that are involved in scientific activities that address major societal challenges around agriculture, food and the environment is enormous. They range from researchers working across geographical areas and scientific domains, to policy makers designing development and innovation interventions. These activities have been traditionally informed and powered by a combination of quite heterogeneous data sources and formats, as well as several research infrastructures and facilities, at a local, national, and regional level. In 2010, a SCAR study tried to give an overview of this picture, which has been documented in the report "*Survey on research infrastructures in agri-food research*" [9]. As more and more research information and IT systems became available online, the relevance of agricultural knowledge organisation schemes and systems became higher. A recent foresight paper on the topic has been published by the SCAR working group "*Agricultural Knowledge and innovation systems*" [1]. The emergence of the open access and data policies has brought forward new challenges and opportunities (as the 2015 *GODAN Discussion Paper* has revealed) [8], which have to be addressed and supported by future e-infrastructure services and environments. In addition to this, Commissioner Moedas pointed out a clear link between societal challenges and openness as a European priority. He has positioned these challenges across the three dimensions of Open Science, Open Innovation and Openness to the World. AGINFRA+ is best positioned to achieve impact upon all these dimensions.

6.5.1 Open Science

The Belmont Forum[1] is a roundtable of the world's major funding agencies of global environmental change research and international science councils, which collectively work on how they may address the challenges and opportunities associated with global environmental change. In 2013, the Belmont Forum initiated the multi-phased *E-Infrastructures and Data Management* Collaborative Research Action[2]. In August 2015, this initiative published its recommendations on how the Belmont Forum can leverage existing resources and investments to foster global coordination in e-Infrastructures and data management, in a report entitled "A Place to Stand: e-Infrastructures and Data Management for Global Change Research" [2]. The main recommendations of this report included (a) to adopt data principles that establish a global, interoperable e-infrastructure with cost-effective solutions to widen access to data and ensure its proper management and long-term preservation, (b) to promote effective data planning and stewardship in all research funded by Belmont Forum agencies, to enable harmonization of the e-infrastructure data layer, and (c) to determine international and community best practice to inform e-infrastructure policy for all Belmont Forum research, in harmony with evolving research practices and technologies and their interactions, through identification of cross-disciplinary research case studies.

AGINFRA+ is fully aligned with these recommendations and has as its strategic impact goal to ensure that Europe brings forward to the Belmont Forum AGINFRA as a world-class data infrastructure. It aims to demonstrate a number of cost-effective and operational solutions for access, management and preservation of research data that will be based upon (and take advantage of) the core e-infrastructure services at hand. And to introduce, through a number of prototype demonstrators, the three selected research use cases as good practices that may become international and community best practices and drive e-infrastructure policy for all Belmont Forum research.

6.5.2 Open Innovation

The agricultural business landscape is rapidly changing. Established brands in agriculture such as John Deere, Monsanto, and DuPont are now as much data-technology companies[3] as they are makers of equipment and seeds. Even though agriculture has been slower and more cautious to adopt big data than other industries, Silicon Valley and other investors are taking notice. Startups like *Farmers Business Network*[4],

[1]https://www.belmontforum.org

[2]http://www.bfe-inf.org/info/about

[3]http://techcrunch.com/2013/10/02/monsanto-acquires-weather-big-data-company-climate-corporation-for-930m/

[4]https://www.farmersbusinessnetwork.com/

which counts Google Ventures as an investor, have made collecting, aggregating, and analysing data from many farms their primary business. Popular, business and tech press keeps on highlighting the evolution that (big) data brings into the agriculture, food and water business sectors—but also into helping feed 9 billion people [3, 5–7, 10]. For instance, in the farming sector, data collection, management, aggregation and analytics introduce a wide variety of innovative applications and solutions such as, for example, sensors which can tell how effective certain seed and types of fertilizer are in different sections of a farm. Another example could be a software which may instruct the farmer to plant one hybrid in one corner and a different seed in another for optimum yield or intelligent systems which may adjust nitrogen and potassium levels in the soil in different patches. All this information can be automatically also shared with seed companies to improve hybrids.

This is also creating an investment environment with a tremendous potential for startups and companies that are focusing on data-intensive applications. Last year's investment report from AgFunder[5] states that the $4.6 billion that was raised by agricultural technology (AgTech) companies in 2015 from private investors, is nearly double 2014's $2.36 billion total—and outpaced average growth in the broader venture capital market. It also points out that this should not be considered as a new *tech bubble*: apart from the food e-commerce sector that seems overheated, the rest of the agriculture and food market is still facing challenges not seen in other mainstream technology sectors. In comparison to the size of the global agriculture market ($7.8 trillion), AgTech investment (with less than 0.5% of it) seems to be very modest and with amazing prospects.

AGINFRA+ particularly aims to take advantage of this investment trend by targeting and involving agriculture and food data-powered companies (and especially startups and SMEs). It has a dedicated work activity on getting such companies involved, and it will align its efforts with the business outreach (through data challenges, hackathons, incubators and accelerators) of its European (e.g. ODI[6], Big Data Value Association[7]) and global networks (e.g. GODAN[8]).

6.5.3 Openness to the World

At the 2012 G-8 Summit, G-8 leaders committed to the *New Alliance for Food Security and Nutrition*, the next phase of a shared commitment to achieving global food security. As part of this commitment, they agreed to "*share relevant agricultural data available from G-8 countries with African partners and convene an international conference on Open Data for Agriculture, to develop options for the establishment of a global platform to make reliable agricultural and related information available to*

[5]https://agfunder.com

[6]http://theodi.org/

[7]http://www.bdva.eu/

[8]http://godan.info/

African farmers, researchers and policymakers, taking into account existing agricultural data systems" [12]. In April 2013, the prestigious G-8 International Conference on Open Data for Agriculture took place in Washington DC, announcing the G8 Open Data Action plans[9]. The goal of the EC's action plan[10] has been "Open access to publicly funded agriculturally relevant data" and included the flagship projects (such as agINFRA, SemaGrow, TRANSPLANT, OpenAIRE) that some of the key AGINFRA+ partners were coordinating and implementing at that time. To a large extend, when the *Global Open Data for Agriculture and Nutrition (GODAN)* initiative was launched as a result of this conference, these EC-funded projects have been driving the contributions of partners like ALTERRA and Agroknow that made Europe one of the GODAN global leaders.

In a similar way, and through its representation and active contribution to international networks like *GODAN*, the *Research Data Alliance (RDA)* and the *Coherence in Information for Agricultural Research and Development (CIARD*[11]*)*, AGINFRA+ aims to continue supporting the global outreach and collaboration of European agriculture and food data stakeholders with their international counterparts.

6.6 Conclusions and Future Work

The proposed e-infrastructure addresses the challenge of supporting user-driven design and prototyping of innovative e-infrastructure services and applications. It particularly tries to meet the needs of the scientific and technological communities that work on the multi-disciplinary and multi-domain problems related to agriculture and food. It will use, adapt and evolve existing open e-infrastructure resources and services, in order to demonstrate how fast prototyping and development of innovative data- and computing-intensive applications can take place.

In order to realize its vision, AGINFRA+ will achieve the following objectives:

- identify the requirements of the specific scientific and technical communities working in the targeted areas, abstracting (wherever possible) to new AGINFRA services that can serve all users;
- design and implement components that serve such requirements, by exploiting, adapting and extending existing open e-infrastructures (namely, OpenAIRE, EUDAT, EGI, and D4Science), where required;
- define or extend standards facilitating interoperability, reuse, and repurposing of components in the wider context of AGINFRA;
- establish mechanisms for documenting and sharing data, mathematical models, methods and components for the selected application areas, in ways that allow their discovery and reuse within and across AGINFRA and served software applications;

---

[9]https://sites.google.com/site/g8opendataconference/action-plans

[10]https://docs.google.com/file/d/0B4aXVC8hUc3oZlVEdlZ1RVJvZms/edit

[11]http://ciard.info/

- increase the number of stakeholders, innovators and SMEs aware of AGINFRA services through domain-specific demonstration and dissemination activities.

Furthermore, AGINFRA+ will focus on the development of fully defined demonstrator applications in three critical application areas, which will allow to showcase and evaluate the infrastructure and its components in the context of specific end-user requirements from different scientific areas.

Acknowledgements The work presented in this chapter has been partly supported by the AGINFRAPLUS Project that is funded by the European Commission's Horizon 2020 research and innovation programme under grant agreement No 731001

References

1. S.W. AKIS-3, Agricultural knowledge and information systems towards the future—a foresight paper (2016). Retrieved from http://ec.europa.eu/research/scar/pdf/akis-3_end_report.pdf#view=fit&pagemode=none
2. Belmont Forum, A place to stand: e-infrastructures and data management for global change research (2015). Retrieved from http://www.bfe-inf.org/sites/default/files/A_Place_to_Stand-Belmont_Forum_E-Infrastructures__Data_Management_CSIP.pdf
3. D. Bobkoff, Seed by seed, acre by acre, big data is taking over the farm (2015, September 15). Retrieved from http://www.businessinsider.com/big-data-and-farming-2015–8
4. K. Bronson, I. Knezevic, Big data in food and agriculture. Big Data and Society **2016**, 1–5 (2016)
5. C. Fishman, Water is broken. data can fix it (2016, March 17). Retrieved from https://www.nytimes.com/2016/03/17/opinion/the-water-data-drought.html?_r=0
6. L. Gilpin, How big data is going to help feed nine billion people by 2050 (2016). Retrieved from http://www.techrepublic.com/article/how-big-data-is-going-to-help-feed-9-billion-people-by-2050/
7. C. Metz, Forget GMOs. The future of food is data—mountains of it (2014). Retrieved from https://www.wired.com/2014/09/ex-googler-using-big-data-model-creation-new-foods/
8. Open Data Institute, How can we improve agriculture, food and nutrition with open data (2015). Retrieved from http://www.godan.info/sites/default/files/old/2015/04/ODI-GODAN-paper-27-05-20152.pdf
9. Survey on Research Infrastructures in Agri-food Research (2010). Retrieved from https://ec.europa.eu/research/scar/pdf/final_scar_survey_report_on_infrastructures.pdf
10. A. Thusoo, How big data is revolutionizing the food industry (2014). Retrieved from https://www.wired.com/insights/2014/02/big-data-revolutionizing-food-industry/
11. F.K. van Evert, S. Fountas, D. Jakovetic, V. Crnojevic, I. Travlos, C. Kempenaar, Big data for weed control and crop protection. Weed Res. **57**, 218–233 (2017)
12. White House Office of The Press Secretary, Fact sheet: G-8 action on food security and nutrition (2012, May 18). Retrieved from https://obamawhitehouse.archives.gov/the-press-office/2012/05/18/fact-sheet-g-8-action-food-security-and-nutrition
13. S. Wolfert, L. Ge, C. Verdouw, M.-J. Bogaardt, Big data in smart farming—a review. Agric. Syst. **153**, 69–80 (2017)

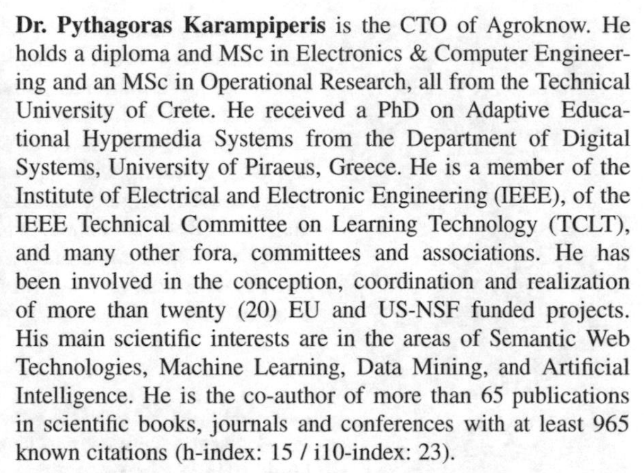

Dr. Pythagoras Karampiperis is the CTO of Agroknow. He holds a diploma and MSc in Electronics & Computer Engineering and an MSc in Operational Research, all from the Technical University of Crete. He received a PhD on Adaptive Educational Hypermedia Systems from the Department of Digital Systems, University of Piraeus, Greece. He is a member of the Institute of Electrical and Electronic Engineering (IEEE), of the IEEE Technical Committee on Learning Technology (TCLT), and many other fora, committees and associations. He has been involved in the conception, coordination and realization of more than twenty (20) EU and US-NSF funded projects. His main scientific interests are in the areas of Semantic Web Technologies, Machine Learning, Data Mining, and Artificial Intelligence. He is the co-author of more than 65 publications in scientific books, journals and conferences with at least 965 known citations (h-index: 15 / i10-index: 23).

Ir. Rob Lokers is a project manager of ICT projects and research infrastructures at Alterra, Wageningen UR. He has a long year experience in information system development in scientific and commercial environments. His expertise is in ICT in science (agriculture, food security, environment), research infrastructures, knowledge technology and agro- environmental modelling. Rob's focus is on co-creating innovative knowledge infrastructures and applications that advance the level of agro-environmental research by exploiting the possibilities of Big Data and the Semantic Web. He leads projects and work packages in large international research projects and service contracts, e.g. research infrastructure initiatives like the FP7 SemaGrow and Trees4Future projects and knowledge platforms like Climate-ADAPT, the European Platform for Climate Adaptation. Rob is an IPMA, level C certified project manager.

Dr. Pascal Neveu is the Director of the MISTEA Joint Research Unit since 2009 and is the Leader of the CODEX Centre. His work is focused on the development of data integration and computer science methods devoted to analysis and decision support for Agronomy and Environment, with particular emphasis on modelling, dynamical systems and complex systems. He is leader of the PHENOME MCP2 collaborative Project.

Mrs. Odile Hologne is the head of the department of scientific information at the French institute for agricultural research. For many years, she has managed projects and teams linked to information technologies in the agri-food sciences. In her current position she is involved in many international working groups dealing with "open science in the agri-food sector" such as GODAN (global open data for agriculture and nutrition) and RDA (Research data alliance). She is one of the associated expert of RDA Europe and she is the coordinator the project eROSA «Towards an e-infrastructure roadmap for open science in agriculture» (www.erosa.aginfra.eu)

Mr. Georgios Kakaletris is a graduate of the Department of Physics at the National and Kapodistrian University of Athens, holds a M.Sc. in Informatics and is a PhD candidate in the Department of Informatics and Telecommunications, of the same institution. He has worked for major Greek companies (Profile SA., Intelltech SA/Quest SA., Decision SA), initially as an engineer and software designer (S/W Engineer-Architect), then as Project Manager (IT Project Manager) and during the last years as Director of the Research & Development Department. He has been acting as an IT consultant for many private and public bodies such as the National and Kapodistrian University of Athens, the Greek Parliament etc. His experience covers the following sectors: Design, development and implementation of integrated information systems for both the public and private sectors, Management of information technology projects, Real-time data transmission services, Large scale financial applications, Data management systems, Distributed Systems etc.

Dr. Leonardo Candela is a researcher at Networked Multimedia Information Systems (NeMIS) Laboratory of the National Research Council of Italy - Institute of Information Science and Technologies (CNR - ISTI). He graduated in Computer Science in 2001 at University of Pisa and completed a PhD in Information Engineering in 2006 at University of Pisa. His research interests include Data Infrastructures, Virtual Research Environments, Data Publishing, Open Science, Scholarly Communication, Digital Library [Management] Systems and Architectures, Digital Libraries Models, Distributed Information Retrieval, and Grid and Cloud Computing. He has been involved in several EU-funded projects. He was a member of the DELOS Reference Model Technical Committee and of the OAI-ORE Liaison Group.

Mr. Matthias Filter is senior research scientist in the unit "Food Technologies, Supply Chains and Food Defense" at the Federal Institute for Risk Assessment (BfR), Germany. Matthias coordinated several software development and modelling activities within the BfR. In several national and international research projects Matthias promoted the idea of standardized information exchange as the basis for harmonization, transparency and knowledge generation. Matthias holds a diploma in Biochemistry and has been working as research scientist and project manager in public and private sector organizations for more than 10 years. He has been appointed as external advisor for EFSA and several EU research projects.

Dr. Nikos Manouselis is a co-founder and the CEO of Agroknow. He has a diploma in Electronics & Computer Engineering, a M.Sc. in Operational Research, and a M.Sc. in Electronics & Computer Engineering, all from the Technical University of Crete, Greece. He also holds a PhD on agricultural information management, from the Agricultural University of Athens (AUA), Greece. He has extensive experience in designing and coordinating large scale initiatives related to the organization, sharing, and discovery of agricultural information for research, education and innovation. He is a member of the Steering Group of the AGRIS service of the Food & Agriculture Organization (FAO). He has contributed to the conception, coordination, and implementation of several projects that have been supported with funding from the European Commission - such as Organic.Edunet, CIP PSP VOA3R, CIP PSP Organic.Lingua, FP7 agINFRA, FP7 SemaGrow, H2020 OpenMinTed and H2020 Big Data Europe.

Dr. Maritina Stavrakaki is an Agriculturist and holds a diploma (BSc) in Crop Science and Engineering, a MSc in Viticulture and Enology, and a PhD in Viticulture, all from the Agricultural University of Athens (AUA), Greece. She is involved in the use of ampelographic and molecular methods and tools for the identification and discrimination of grapevine varieties and rootstock varieties, as well as in the morphology, physiology and eco-physiology of the vine. Over the last few years, she has participated in various scientific programs with the Laboratory of Viticulture. Her previous experience includes teaching Viticulture both at the Agricultural University of Athens, Greece and at the National and Kapodistrian University of Athens, Greece and working as an agriculturist at the Hellenic Ministry of Rural Development and Food. She has participated in many scientific congresses and symposia with presentations and publications.

Dr. Panagiotis Zervas is the Director of Project Management at Agroknow, since April 2016. He holds a Diploma in Electronics and Computer Engineering from the Technical University of Crete, Greece in 2002, a Master's Degree in Computational Science from the Department of Informatics and Telecommunications of the National and Kapodistrian University of Athens, Greece in 2004 and a PhD from the Department of Digital Systems, University of Piraeus, Greece in 2014. He has 17-year professional experience in the conception, coordination and implementation of 25 R&D Projects. He is co-chair of the Working Group on Agrisemantics of the Research Data Alliance (RDA). He has also been involved in the design and implementation of large-scale EU-funded initiatives such as CIP Inspiring Science Education and FP7 Go-Lab. He is the co-author of more than 90 scientific publications with more than 680 citations (h-index: 14 / i10-index: 27), as listed in Scholar Google.

Chapter 7
Green Neighbourhoods: The Role of Big Data in Low Voltage Networks' Planning

Danica Vukadinović Greetham and Laura Hattam

Abstract In this chapter, we aim to illustrate the benefits of data collection and analysis to the maintenance and planning of current and future low voltage networks. To start with, we present several recently developed methods based on graph theory and agent-based modelling for analysis and short- and long-term prediction of individual households electric energy demand. We show how maximum weighted perfect matching in bipartite graphs can be used for short-term forecasts, and then review recent research developments of this method that allow applications on very large datasets. Based on known individual profiles, we then review agent-based modelling techniques for uptake of low carbon technologies taking into account socio-demographic characteristics of local neighbourhoods. While these techniques are relatively easily scalable, measuring the uncertainty of their results is more challenging. We present confidence bounds that allow us to measure uncertainty of the uptake based on different scenarios. Finally, two case-studies are reported, describing applications of these techniques to energy modelling on a real low-voltage network in Bracknell, UK. These studies show how applying agent-based modelling to large collected datasets can create added value through more efficient energy usage. Big data analytics of supply and demand can contribute to a better use of renewable sources resulting in more reliable, cheaper energy and cut our carbon emissions at the same time.

7.1 Introduction

Our planet is heavily influenced by how electric energy is generated. Electricity comes mostly from fossil fuels, nuclear fuels and renewable resources. Both globally

---

D. V. Greetham (✉) · L. Hattam
Department of Mathematics and Statistics, Centre for the Mathematics of Human Behaviour, University of Reading, Reading RG6 6AX, UK
e-mail: d.v.greetham@reading.ac.uk

L. Hattam
e-mail: l.hattam@reading.ac.uk

A. Emrouznejad and V. Charles (eds.), *Big Data for the Greater Good*, Studies in Big Data 42, https://doi.org/10.1007/978-3-319-93061-9_7

and in the UK, the energy mosaic is complex, there are many important factors that will reshape the production and use of electric energy in the near future.

Among them are: the mass roll-out of energy smart meters expected in 2020 in the UK [19]; bigger percentages of renewable generation in national energy supply, especially solar and wind [37]; predicted electrification of transport and heating through electric vehicles and heat pumps [41]; more affordable energy storage [42], [13] and many others. The listed factors also mutually influence each other. For example, the combination of solar panels and energy storage [32] might make the uptake of both happen sooner than previously expected. All this results in a complex system difficult to model and predict.

In the last two decades, half-hourly or higher resolution measurements of individual households and commercial properties demand became available through different pilot projects [7, 31]. This resulted in more understanding in individual household energy demand and how different new technologies impact this demand. There is still a lot to be done in understanding energy use on individual level and creating accurate forecasts that can be used in so called *smart grid*. This includes distributed generation and so called *prosumage* where the end-users, traditionally only con*sum*ers, are now also involved in *pro*duction and stor*age* of energy. Some other aspects of smart grid are adaptive control of devices, automatic demand response, optimal use of secondary generating sources and optimised charging and discharging of storage. For all of those aspects, efficient solutions cannot be obtained without big data collection and analysis.

Another issue that we encounter and where big data might help is that the energy objectives on the national level, where one is looking at the aggregated demand, and on the local level, where one is concerned with low voltage networks infrastructure, might be conflicting. For example, regionally or nationally, it might make sense to rectify the usual late afternoon demand peak by incentivising customers to charge their electric vehicles (EVs) during the night. On the other hand, uncontrolled applications of time of use tariffs could be extremely detrimental to low voltage networks that were designed with behavioural diversity taken into account. In some cases, where there is a fast uptake of EVs concentrated in few neighbourhoods, local networks would not be able to cope with sudden night peaks without costly reinforcements. In these circumstances high resolution data can help by enabling modelling and constrained optimisation on a local level that can be scaled up to asses how the decisions made locally would impact the aggregated demand. Also, data analysis can help to react quickly in the cases where proposed policies would create inefficiencies or damage some of the stake-holders as in the example above.

7.2 Short-Term Forecasting of Individual Electric Demand

Short-term forecasting of energy demand is usually taken to be forecasting one day up to one-two weeks in the future [21]. While distribution network/system operators were doing aggregated demand forecasts for more than 60 years, the individual level forecasting is newer and came about when the individual data in higher resolution first became available.

7.2.1 Methodology

As we mentioned before, load forecasting was traditionally done at the medium voltage (MV) to high voltage (HV) networks level. At that level of aggregation, the demand is more smooth and regular [45]. There is a plethora of methods for short-term forecast of aggregated demands, mostly coming from artificial intelligence [48](such as artificial neural networks, fuzzy logic, genetic algorithms, expert systems and similar) and statistics [46] (linear regression, exponential smoothing, stochastic time series, autoregressive integrated moving average models (ARIMA), etc.). More recently, probabilistic methods allowing for ensemble forecasts are being developed [21].

On the other hand, methods for short term individual demand forecast are somewhat scarcer. Historically, only quarterly meter readings were available, and distribution network operators were more concerned about commercial (large) customers. This started to change when the half-hourly or higher resolution data became available and when new technologies such as EVs and photovoltaics became a cause of concern for future in distribution networks, because of increased demand but also because of issues with voltage, thermal constraints, network balance etc. [8, 9, 11, 39]. At the LV network level (from an individual household to a feeder or substation), the demand is less regular and normally has frequent but irregular peaks [3]. Therefore new methods need to be developed that work well for this kind of inputs.

7.2.2 Recognised Issues

Several issues are recognised in the literature when predicting the individual level demand. Firstly, errors are normally much larger than with aggregated demands. In [9] it is claimed that aggregated forecast errors are in brackets of 1–2% when Mean Absolute Percentage Error (MAPE) is considered, while at the individual level errors raise to 20% and higher. Secondly, in [39] it has been noted that while most forecasts concentrate on the mean load, they actually have significant errors predicting peaks. Predicting peaks accurately, on the other hand, is of utmost importance for distribution network (system) operators for all kind of applications: demand side response, efficient use of battery storage etc. Thirdly, it was noticed in [12] that on individual level, demand is influenced by many different types of behaviour resulting in frequent but irregular peaks, and therefore volatile and noisy data.

7.2.3 Forecasts Quality - Measuring Errors

Above mentioned issues combined bring us to the problem of measuring quality of different individual short-term forecasts. If we denote with y_t an observed demand at time t and f_t is a forecast of y_t then the forecast error is commonly defined as $e_t = y_t - f_t$ [23]. Mean square error $MSE = \text{mean}(e_t^2)$ and its variants, root mean square error, mean absolute error etc, are all examples of *scaled* errors. That means that their scale depends on the scale of data [23]. The percentage error is given by $p_t = \frac{100e_t}{y_t}$ and it is used to define scale independent errors, such as mean absolute percentage error $MAPE = \text{mean}(|p_t|)$, and other variants. It was noted in [39] that predicting peaks is very challenging. For that reason in operational settings, variants of MSE are sometimes replaced with $p-$norms, thus taking higher powers of p (e.g. $p = 4$ instead of $p = 2$) in order to highlight peaks amplitudes. So, errors are calculated as $\sqrt[p]{\sum_t e_t^p}$, for some $p \in \mathbb{N}$.

7.2.4 Adjusted Errors

For noisy and volatile data, a forecast that predicts peaks accurately in terms of size and amplitude, but displaced in time, will be doubly penalised, once for predicted and once for missed peak. Thus peaky forecasts can be easily outperformed by smooth forecasts, if using point-wise error measures.

In [12] the authors suggested an adjusted error measure that allows slight displacements in time. This displacement in time can be reduced to an optimisation problem in bipartite graphs, namely the minimum weight perfect matching in bipartite graphs [36] that can be solved in polynomial time. A graph $G = (V, E)$ is bipartite if its vertices can be split into two classes, so that all edges are in between different classes. In this case observations y_t and forecasts f_t make two natural bipartite classes. Errors between observations and forecasts are used as weights on the edges between the two classes. Instead of just looking at errors $e_t = y_t - f_t$ (i.e. solely considering the edges between y_t and f_t), also taken into account are differences between

$$y_t - f_{t-1}, y_t - f_{t+1}, y_t - f_{t-2}, y_t - f_{t+2}, \ldots, y_t - f_{t-w}, y_t - f_{t+w},$$

for some time-window w. These differences are added as weights and some very large number is assigned as the weight of all the other possible edges between two classes, in order to stop permutations of points far away in time. Now, we want to find the perfect matching that minimises the sum of all weights, therefore allowing possibility of slightly early or late forecasted peaks to be matched to the observations without the double penalty. Using errors as weights we can solve this as the minimum weighted perfect matching in polynomial time [29], but also allow localised permutations in forecasts without large computational cost as in [5]. The Hungarian algorithm with a time complexity of $O(n(m + nlogn))$ for graphs with n nodes and m edges [29,

47] when n is quite large, for example when forecasting in very high resolution, can be replaced by much faster suboptimal alternatives that use parallelisation. The auction algorithm [35] is able to find matchings comparable with ones produced by Hungarian algorithm, but is simpler to implement and much faster [20] for large n.

7.2.5 Short-Term Forecasts Based on Adjusted Errors

The idea of an adjusted error resulted in two new forecasts that allow for the 'peakiness'. This can be useful for electricity network distribution operators for different demand side response and storage dynamic control applications.

A new forecast, *'adjusted average'*, attempting to preserve 'peakiness' by allowing slight shifts in time in the historical data is presented in [12]. Given N historical profiles (for example N half-hourly demands during previous k Thursdays) the aim is to predict demand for coming Thursday. The initial baseline F^1 can be defined as a median of N profiles (or similar). Then, the kth historical demand profile $G(k)$ is matched through constrained permutations with the current baseline so that nearby peaks (e.g. an hour and a half in past and future) match up and in this way a new matched profile $\hat{G}^k$ is obtained. The new baseline is then calculated as

$$F^{k+1} = \frac{1}{k+1}(\hat{G}^k + kF^k). \quad (7.1)$$

This process is repeated k times and the final forecast is

$$F^N = \frac{1}{N+1}(\sum_{k=1}^{N} \hat{G}^k + F^1). \quad (7.2)$$

Another forecast, *'permutation merge'* is presented in [5]. Starting from the same premise as *'adjusted average'*, instead of averaging baseline and a matched profile in each step to obtain the new forecast, here a profile is chosen instead that minimises the sum of the adjusted errors in each step. This results in a faster forecast when allowed permutations are close to the forecasted point, i.e. a permutation window is small[1].

On Fig. 7.1 adjusted average and permutation merge forecast were compared (the benchmark was mean forecast that, based on k historical profiles for a day in question, e.g. Tuesday, forecasts each half hour of that day as a mean of k previous Tuesdays).

The data was Irish Smart meter data [7]. The adjusted error measure (with $w = 3$ and $p = 4$, see Sect. 7.2.2 and 7.2.4) was used to calculate the errors. The PM forecast improved as more historical profiles were added, and was also much faster than adjusted average on this dataset for all $w < 6$.

[1]For example, for half-hourly data if predicted peaks are early/late up to one hour and a half that means that a permutation window is 3.

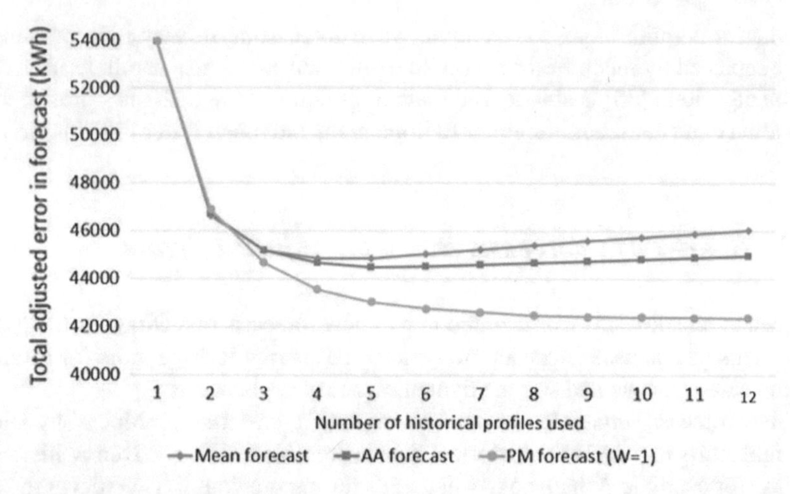

Fig. 7.1 A comparison of adjusted average, permutation merge and mean forecast. From [5]

7.2.6 *Applications*

Accurate short-term forecasts of individual demand are important for distribution network and systems operators (DNO, DSO) for efficient planning and managing of low voltage networks.

Several important applications arise through *demand side response* (DSR), a set of measures that helps DNOs to manage load and voltage profiles of the network. Under this regime, consumers are incentivised to lower or shift their electricity use at peak times. For example, Time-of-Use tariffs can be implemented in order to reduce aggregated demand peaks (such as well-known early evening peak for instance). Individual forecasts then allow network modellers to predict consumers reactions on tariffs [1] and to avoid undesired effects such as move of the excessive demand overnight or a loss of natural diversity of demand that could potentially exacerbate peaks instead of mitigating them. Another obvious application is in control of small energy storage (at household or residential level).

Set point control is the simplest and most common way of controlling battery storage, but its simplicity is paid by inefficiency as it rarely succeeds in reducing the peak demand. Efficiency can be improved significantly through accurate household-level forecasts and dynamic control. This would help to optimally control charging and discharging of batteries [27, 34].

7.3 Long-Term Forecasting: Modelling Uptake of Low Carbon Technologies

As more data is becoming available on individual households' level energy consumption and low carbon technologies, such as electric vehicles, photovoltaics and heat pumps, we are facing new challenges and opportunities.

The individual and combined impact of different low carbon technologies on electric networks presents us with numerous challenges for accurate mathematical modelling and forecasting. Again, several inter-dependent factors come into play: tariffs, socio-demographics, occupancy, life styles, etc. (see [22, 26, 30, 49]). The time of use will differ for different technologies - while photovoltaics are physically limited to a certain period of day, the use of an electric car will be determined by social and environmental patterns. Considering the availability of electric storage and vehicle to grid technologies, we might become more flexible. However, to capture this behaviour, quite complex approaches to modelling, forecasting and optimisation problems are needed.

Uptake of low carbon technologies can be modelled as innovation diffusion. One of the most influential early studies on innovation diffusion within a social system was provided by [33]. The author suggested that the spread of innovations over time resulted from individuals belonging to a social network communicating with one another.

This theory has been further developed from a top-down (macroscopic) and a bottom-up (microscopic) perspective, which can both be applied to predict the adoption of low-carbon technologies (LCTs) within a sample population. A macroscopic approach uses differential equations to determine the diffusion process, whereas, a microscopic viewpoint applies agent-based modelling techniques.

7.3.1 Macroscopic Approaches - Differential Equation Models

Initially, it was aggregate models that followed the workings of [33]. In the seminal work [2], Bass proposed the following nonlinear ordinary differential equation to govern the adoption process

$$\frac{dF}{dT} = (p + qF)(1 - F), \tag{7.3}$$

where F is the adopter ratio, T is time, p is the innovation coefficient and q is the imitation coefficient. The equation parameters p and q correspond to external and internal effects respectively such that innovators adopt independent of their social interactions, whilst imitators are influenced by these exchanges. Example external pressures include government incentives and advertising.

From (7.3), the solution is then derived as

$$F(T) = \frac{1 - \exp(-(p + q)T)}{(q/p)\exp(-(p + q)T) + 1}, \tag{7.4}$$

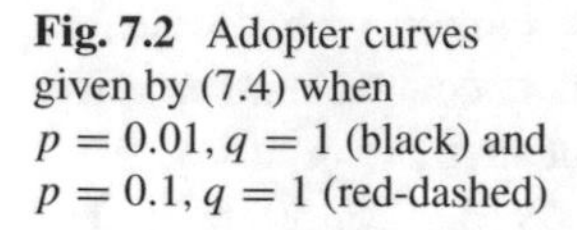

Fig. 7.2 Adopter curves given by (7.4) when $p = 0.01, q = 1$ (black) and $p = 0.1, q = 1$ (red-dashed)

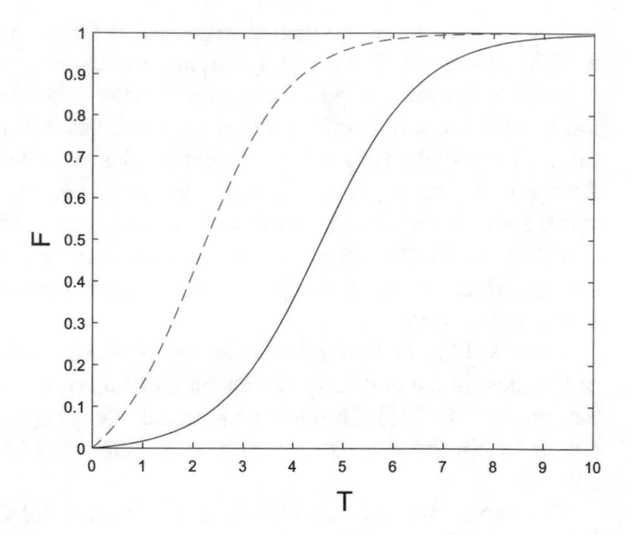

which is plotted in Fig. 7.2 for $p = 0.01, q = 1$ (black) and $p = 0.1, q = 1$ (red-dashed). These are S-shaped adopter curves, which describe the temporal dependence of the adoption behaviour and how this varies with p and q.

Later, this model was extended in [18] by also including spatial variables with a diffusion coefficient. As a result, innovations propagate over time and space, like a wave. This comparison to wave motion was made by [28], where real-life data was used.

The equation that was developed by [18] can be used to forecast the uptake of LCTs across specific geographies. For instance, [38] applied this nonlinear system to estimate the adoption behaviour for hybrid cars within Japan. The innovation, imitation and diffusion coefficients were chosen by fitting the overall adopter curve to real information. Karakaya [24] predicted the uptake of photovoltaics within southern Germany, although they focussed on the diffusive effects, excluding the innovation and imitation influences, which meant that instead, the heat equation was used. A semi-hypothetical case was presented, where the diffusive coefficient was modified over the domain of interest so to reflect a neighbourhood's likelihood of adoption.

7.3.2 Agent-Based Modelling

Aggregate models are informative tools for studying the overall dynamics, however sometimes they fail to capture the more intricate interactions of individuals within a social system. Consequently, agent-based models (ABMs) have been also developed to determine innovation diffusion. These can characterise an agent's decision process whilst obtaining a macroscopic understanding by aggregating the agents' behaviours. An early ABM was proposed by [6], where individuals were assigned a probability of adoption that corresponded to their awareness of the innovation price and performance. Starting at a micro level, they then examined the subsequent dynamics from an aggregate perspective, and therefore, conducting a bottom-up approach.

In [25] the authors concluded that by assuming agent i has the probability of innovation uptake

$$P = \left(p + \frac{\sum_{i=1,\ldots,M} x_i}{M} q \right) (1 - x_i), \tag{7.5}$$

where $x_i = 1$ if agent i has adopted and $x_i = 0$ if not, and M is the total number of agents, then the behaviour observed overall is consistent with the Bass model. More specifically, the aggregate adopter trajectories that result are of the same form as the S-curves derived by Bass and depicted in Fig. 7.2. Moreover, they are also dependent upon the external and internal parameters, p and q respectively, which were first introduced by [2].

More recently, [10] outlined an agent-based technique for calculating an individual's likelihood of technology adoption, where the overall behaviour was determined with multiple numerical simulations. This model incorporated the effects of advertising, social interactions and a resistance to change at a microscopic level. In addition, agents could reverse their decision of uptake and revert to using the original technology. This unique model feature led to interesting adoption dynamics, including cyclical uptake results.

7.3.3 Hybrid Models

Recent developments in the field of innovation diffusion have combined the bottom-up and top-down approaches. These models are referred to as hybrid simulations. For instance, [44] forecasted the spread of technologies across the world by applying a hybrid model. More specifically, the theory of [2] determined the adoption timings for various groups of countries. Then, agent-based techniques were employed to model the behaviour of individual countries, which were influenced by neighbouring countries and their own preferences.

Another example of a hybrid technique was proposed by [16]. This study focussed upon a local electricity network located in the UK that consisted of multiple feeders, where households were linked to the network by their feeder. The adoption process at a macro-level was governed by the equation derived by [18]. Initially, the network was depicted with a finite element grid and then the finite element method was applied to model innovation diffusion with Haynes' equation. This resulted in feeder uptake curves as a function of time being identified. By following these estimated feeder trends, the adoption behaviour at a household level was next simulated.

Hence, the macro and micro level methods are viable forecasting tools for the uptake of technologies across a sample population. More importantly, the merging of the two approaches provides many varied and complex solutions to the problem of innovation diffusion modelling. An overview and guide for this simulation technique is detailed by [43], where three different types of hybrid models are highlighted and various example studies are described.

7.4 Case Studies

Now we present two case studies of above mentioned techniques used with big data obtained from smart meters.

7.4.1 Modelling Uncertainty - Uptake of Low Carbon Technologies in Bracknell

As a part of the Thames Valley Vision project (www.thamesvalleyvision.co.uk), a realistic low voltage (LV) network was modelled using agent-based modelling. The LV network is situated in Bracknell, UK, comprising of around 600 substations. In this case study, we considered 1841 properties/44 feeders, where 7 were households with PVs installed and 71 were non-domestic properties. The household count connected to each feeder varied, where the minimum number of households along one feeder was 4 and the maximum was 114. Note that here each feeder corresponded to one neighbourhood and all households along a particular feeder were considered neighbours i.e. they were not necessarily adjacent properties. Data sets for 17 selected days (representatives of different seasons, weekdays, weekends and bank holidays) were created using metered data from this LV network. The data sets consisted of a combination of metered and predicted daily demand energy profiles (kWh) for every household, where a genetic algorithm was used to allocate monitored endpoints to unmonitored customers. These profiles had readings every half hour and therefore for each household a load profile with 48 data points was provided. These data sets were referred to as *baseloads*.

The combined scenario of EV and PV uptake was considered. In this model, an agent corresponded to a household. In particular, agents were presumed to be more likely to adopt if their neighbour had a LCT. To analyse the result from a macroscopic viewpoint, multiple runs were undertaken and then the distributions of LCTs were compared at a neighbourhood level. The following outlines the clustering algorithm applied to forecast LCT uptake: firstly, the percentage of households in the sample population that will adopt LCTs and the number of years it will take is established (here, the time horizon was 8 years to simulate the current distribution network operator's regulatory planning period). An initial distribution of LCT seeds was performed. Council tax band (CTB) information and existing PV properties were used to inform the initial seed distribution. The assumption was that larger homes corresponded to higher CTBs and therefore, CTBs were used to identify neighbourhoods with a higher proportion of larger properties.

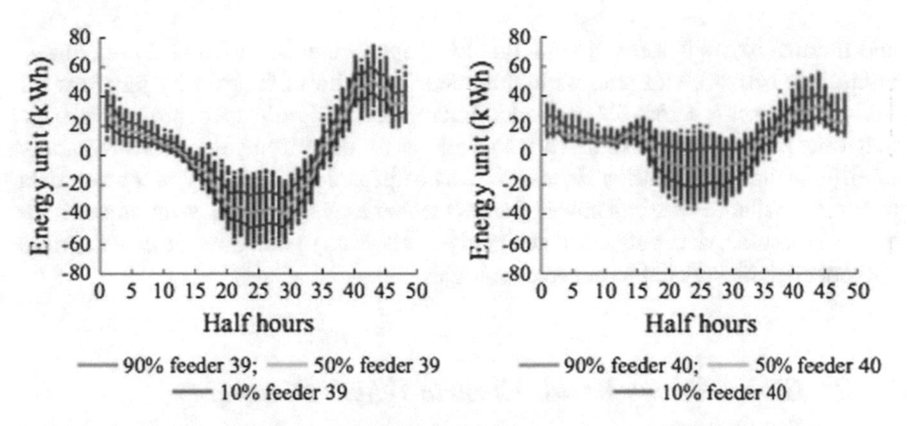

Fig. 7.3 Result of 500 simulations with 30% EV uptake and 30% PV uptake in summer, on two feeders 39 and 40 (seed allocation informed by CTB). From [15]

Then, during the remaining years, LCTs were assigned to households according to the fitness score - this assured that agents with more neighbours who adopted LCT in previous rounds will adopt LCT with higher probability than the agents with none or few neighbours with LCT. The number of LCT households (households that adopted a LCT) increased linearly every year until the specified amount was attained. This allowed to compute impact for different percentages of LCT uptake.

The details of the algorithm can be found in [15]. Once the household obtained LCT, an LCT profile was added (or subtracted in the case of PVs) to its baseload. LCT profile was chosen uniformly at random from the set of available profiles, respecting the seasonal rule: if the baseload applied was representative of spring, summer or winter then the LCT profile chosen also corresponded to spring, summer or winter respectively.

Multiple runs of the model are performed, in order to capture the model uncertainty. The aggregate feeder load resulting from each run is recorded so that eventually a distribution forms at each feeder site. Then, the 10 and 90% quantile are calculated. This gives *confidence bounds* [15], a lower and upper bound for the model response that indicate a minimal and maximal possible daily load. The quantile with the baseload subtracted represents the variation in LCT load at the feeder. Then, dividing the quantiles by the number of households along the feeder, we can compare all the feeders and their LCT loads.

On Fig. 7.3 one can see results obtained for two feeders (feeder 39 with 82 properties, and among those 54% with CTB D or higher on the left and on the left and feeder 40 with 86 properties, and among those 0% with CTB D or higher. Although the number of properties on two feeders is similar and they might be close geographically, there is a notable difference in confidence bounds obtained, resulting in much higher troughs and peaks for feeder 39.

The feeder upper (EVs) and lower bounds (PVs) for the model response can be used as input to a network modelling environment. Consequently, potential voltage and thermal issues that result from this LCT uptake can be assessed. Furthermore, vulnerable network sites can be highlighted. This allows for priority rankings for future investments across LV networks, and represents significant contribution to LV networks planning. This framework also allows for quantifying demand uncertainty of different uptake scenarios. Scenarios can be updated subsequently, when new data becomes available or a trajectory of a new technology uptake takes an unpredicted turn. In this case, data collection and analysis allow us to mitigate some of the risks and uncertainties related to new technologies.

7.4.2 Global Versus Local: Electric Vehicles Charging Incentives

With the accelerated uptake of electric vehicles (EV), significant increase in load is expected in local networks that obtain many vehicles in a short time span.

At national level, electric energy providers would prefer those loads to miss existing afternoon/early evening peak. Therefore, some schemes were piloted were the owners of EVs are incentivised to charge over night. On the other hand, simultaneous charging over night can be detrimental to local networks, as they are designed expecting quieter periods over night.

We used the data from two trials [17], E-mini where the users were incentivised to charge over night, and My Electric Avenue, without such incentive, where the charging was naturally diverse. While this setting has some similarity with electric heating, (for example, in the UK, the existing Economy 7 is a restrictive tariff that encourages household electric storage heating to occur during the night) one must note that the current local networks were designed with Economy 7 in mind, while EV demand is being retrofitted.

Our simulations used the data from a real LV network of 98 feeders in Bracknell. Clustered uptake of EVs was simulated so that households with higher council tax bands had a proportionally higher initial probability to obtain an EV. Social influence was then modelled by a household having a higher probability of uptake in each round of simulation if more neighbours already adopted one. This created clustered uptake through LV. EV load was then added to the households that adopted an EV, using random samples from two trials. Running many simulations allowed us to quantify confidence bounds.

In Fig. 7.4, we calculated a *feeder index*, by taking the maximum value of the 90% feeder quantile (over 500 simulations). The base-load is then subtracted and the result normalised dividing by the number of households along the feeder. In this way we get an index of EV load for each feeder. The blue and red trends relate to the Mini-E trial and the My Electric Avenue trial respectively. One can see that there is a siginificant increase for incentivised night charging.

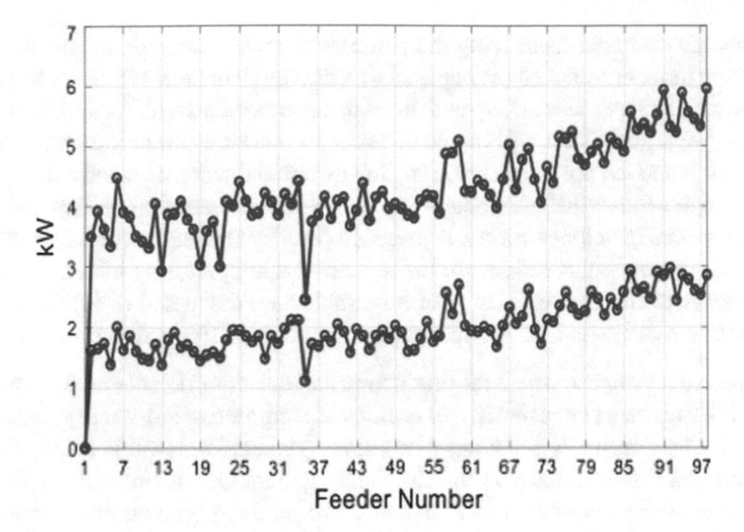

Fig. 7.4 A comparison of the feeder index (for 50% EV uptake). Blue: Mini-E trial, red: My Electric Avenue trial. From [17]

The results [17] have shown that simple incentives that favour charging overnight in order to smooth early evening peak might potentially cause extensive problems for local networks, and that diversity needs to be reintroduced in order to prevent damage of existing local networks in so called green neighbourhoods where the uptake of LCTs is expected to be higher sooner. This is an example where a simple policy solution (incentivising overnight EV charging) that optimises one cost function (reducing evening peak) has detrimental effect on a connected optimisation problem - to maintain LV network at the minimal cost. This just reinforces the fact that a complex system is more than a sum of its components, and needs to be perceived and simulated in its entirety.

7.5 Policy Implications

Energy availability and price plays extremely important role in daily life and subsequently in politics, as we are often reminded. There is an adage that no government can survive continuous power cut.

Big data can help energy policy makers to make more informed decisions, but it should be used in parallel with smaller in-depth studies. This is especially important because of ever-changing technology and human behaviour that interplay and create a complex system.

In [14], based on an in-depth small scale study, the use of smart meters as tools per se for demand reduction is dismissed. It was found that the variability of users demand behaviour is large, confirming the natural diversity observed in larger studies. They also noticed that seeing the patterns of own consumption actually might make them more entrenched (as opposed to reducing demand). While we fully agree that smart meters use cannot help with the reduction demand without additional interventions, we argue that it will be much easier to measure future interventions with smart meters and to spot if they are going in unwanted directions much quicker. Also, methodologies that work from bottom up such as agent-based modelling and uncertainty measures developed for the aggregated profiles will allow for better planning and design of new interventions for the reduction and/or shifting of demand.

The ways to increase efficiency of renewables are discussed in [4]. Three types of policies are discussed, noting that implementations differ across different states:

- net metering policies, where the production and consumption of energy in a household is based on a single meter, and utilities are required to buy back generated, not consumed energy; this encourages owners to generate renewable energy because it decreases financial barriers for distributed generation; again, in our opinion no proper assessments of costs and benefits can be done without smart meter data collection and analysis;
- interconnection standards for distributed generation - this is a more infrastructural point that is not directly connected to big data;
- dynamic pricing policies - here the disadvantage is that households are exposed to price variability, so it is more difficult to e.g. plan an annual budget, but the advantages are that through critical peak pricing and time-of-use tariffs, this can help with reducing total energy price by smoothing peaks and throughs in demand, which again relies critically on big data collection and analysis.

In addition the ownership of smart-metered data and of consumer owned renewables generation is just starting to be regulated, so new rules and regulations could help reduce uncertainties associated with smart-grid investment.

In [40] implications of smart meters installation were discussed and 67 potential benefits and obstacles across technical and socio-economic (vulnerability, poverty, social background, consumer resistance and ambivalence) dimensions are given. They conclude that while the centralised national roll-out through big players assumes a rational consumer with a complete info on single technology at its end, the reality is a very complex system of stakeholders with different needs, and emotional agent that progressively influences the future states of the system.

Both of these issues, the efficiency of renewables and the data ownership are quite complex, and we believe that transparent methods of analysis and public domain data available to create repeatable case-studies can help shedding light to those problems and will result in more efficient and fairer systems for all involved actors.

7.6 Conclusions

In this chapter, we hope that we made our case to show the benefits of data collection and analysis to the maintenance and planning of current low voltage networks and to describe potential future use that would benefit not only distribution system operators, but the whole ecosystem of stakeholders around energy distributed generation, storage and consumption. Beside individual benefits for each end-user, through energy efficiency the greater good is expected also by cutting carbon emissions.

After considering several standard and recently developed methods for analysis and short- and long-term prediction of individual households electric energy demand we reviewed some applications of graph theory that can be used for measuring quality of and producing better short-term forecasts. In particular minimum weighted perfect matching in bipartite graphs is highlighted and recent implementations of this method that allow applications on very large datasets are shown.

We then discussed different modelling techniques for uptake of low carbon technologies starting with equation-based, moving to agent-based modelling and finally reviewing hybrid approaches. While equation-based models use big data for parameter calibration, agent based-models need to take into account socio-demographic characteristics of local neighbourhoods and for this reason big data is even more valuable for those. This includes, besides smart meter data, publicly available data on postcode, incomes, maps, etc. that can significantly improve the accuracy of models' predictions. While some of these techniques are relatively easily scalable, measuring the uncertainty of their results is more challenging. We have presented confidence bounds that allow us to measure uncertainty of the uptake based on different scenarios.

Furthermore, two case-studies are reported, describing applications of these techniques to energy modelling on a real low-voltage network in Bracknell, UK. These studies show how applying agent-based modelling to large collected datasets can create added value through more efficient energy usage. Big data analytics of supply and demand can contribute to a better use of renewable sources resulting in more reliable, cheaper energy and cut our carbon emissions at the same time.

There is a lot of space for improvement on current methodologies. Current limitations include data availability, missing data and data ownership issues which influence models' calibration and validation. This is especially the case in markets like UK, where several different (competing) entities might collect and manage data (even just one type, for example smart meters) for households on the same street. Also a complex system of stakeholders, who range from distributed systems operators through centralised and distributed energy suppliers to the end users makes it difficult to measure fairness and costs/benefits analysis for all of the actors.

In addition, some of the main challenges currently experienced (and expected at least in near future) will be data security, privacy and ownership. In [40] some of the main barriers for data collection are technical glitches of the measuring components (such as smart meters), incompatibility between suppliers (in UK market switching suppliers might mean changing a smart meter too) then security and privacy issues such as hacking and cyber-terrorism, the perception of utilities 'spying' on households etc. and vulnerability issues, such as consumer misunderstanding of smart meters' use, financial burden assuming that the cost is passed to the consumers, health anxieties based on beliefs that WiFi communication of smart meters data back to centre produces non-ionizing electromagnetic fields bad for health and peripheralisation of groups not covered by smart meters roll-out (rural areas, basements and similar) and so on.

We believe that the system efficiency and ways to improvements of current methods will be obtained through a careful balance of all of these factors when collecting the data. More show-cases of benefits brought by analysing the data and transparency with types of data and methods used in analysis will evidently help with this balancing act.

Finally, we lay out some of the future trends. As the data from various sources becomes available, we expect more accurate hybrid and agent-based models to develop, allowing for better spatio-temporal predictions and therefore a better modelling and forecasting input into low voltage network planning and design.

For instance, combining data obtained from Google maps that shows the available surface and orientation for photovoltaics, availability of off-road parking for EV charging and similar with smart meter, high resolution meteorological, satellite and transport data could mitigate current uncertainties around LCT uptake and improve projections of off-grid PV supply locally and nationally.

To conclude, accurate and efficient individual or feeder level short and long term forecasts of electrical energy demand are needed in order to plan for greener future and big data collection and analysis is facilitating further developments in this area.

Acknowledgements This work was supported by Scottish and Southern Electricity Networks through the New Thames Valley Vision Project (SSET203 New Thames Valley Vision), and funded by the Low Carbon Network Fund established by Ofgem.

References

1. S. Arora, J. Taylor, Forecasting electricity smart meter data using conditional kernel density estimation. Omega **59**, 47–59 (2016)
2. F. Bass, New product growth for model consumer durables. Manage. Sci. **15**(5), 215–227 (1969)
3. M. Brabec , O. Rej Konár, E. Pelikán, M. Malý, A nonlinear mixed effects model for the prediction of natural gas consumption by individual customers. Int. J. Forecasting **24**(4), 659–678 (2008). https://doi.org/10.1016/j.ijforecast.2008.08.005. http://www.sciencedirect.com/science/article/pii/S0169207008000976. Energy Forecasting
4. M.A. Brown, Enhancing efficiency and renewables with smart grid technologies and policies. Futures **58**(Supplement C), 21–33 (2014). https://doi.org/10.1016/j.futures.2014.01.001. http://www.sciencedirect.com/science/article/pii/S0016328714000020. SI: Low Carbon Futures
5. N. Charlton, D. Vukadinovic Greetham, C. Singleton, Graph-based algorithms for comparison and prediction of household-level energy use profiles, in *2013 IEEE International Workshop on Inteligent Energy Systems (IWIES)*, 119–124 2013. https://doi.org/10.1109/IWIES.2013.6698572
6. R. Chatterjee, J. Eliashberg, The innovation diffusion process in a heterogeneous population: A micromodeling approach. Manage. Sci. **36**(9), 1057–1079 (1990)
7. Commission for Energy Regulation (CER): Smart metering electricity customer behaviour trials (2007). http://www.ucd.ie/issda/data/commissionforenergyregulation/ [online], data available from Irish Social Science Data Archive

8. R.E. Edwards, J. New, L.E. Parker, Predicting future hourly residential electrical consumption: a machine learning case study. Energy and Buildings **49**, 591–603 (2012). https://doi.org/10.1016/j.enbuild.2012.03.010. http://www.sciencedirect.com/science/article/pii/S0378778812001582
9. K. Gajowniczek, T. abkowski, Electricity forecasting on the individual household level enhanced based on activity patterns. PLOS ONE **12**(4), 1–26 (2017). https://doi.org/10.1371/journal.pone.0174098
10. M. Gordon, M. aguna, S. oncalves, J. glesias, Adoption of innovations with contrarian agents and repentance. Phys.A: Stat. Mech. Appl. **486**(Supplement C), 192–205 (2017)
11. S. Haben, C. Singleton, P. Grindrod, Analysis and clustering of residential customers energy behavioral demand using smart meter data. IEEE Trans. Smart Grid **7**(1), 136–144 (2016). http://centaur.reading.ac.uk/47589/
12. S. Haben, J. Ward, D.V. Greetham, C. Singleton, P. Grindrod, A new error measure for forecasts of household-level, high resolution electrical energy consumption. Int. J. Forecasting **30**(2), 246–256 (2014). https://doi.org/10.1016/j.ijforecast.2013.08.002. http://www.sciencedirect.com/science/article/pii/S0169207013001386
13. P. Hanser, R. Lueken, W. Gorman, J. Mashal, The practicality of distributed PV-battery systems to reduce household grid reliance. Utilities Policy (in press), – (2017). https://doi.org/10.1016/j.jup.2017.03.004. http://www.sciencedirect.com/science/article/pii/S0957178715300758
14. T. Hargreaves, M. Nye, J. Burgess, Keeping energy visible? exploring how householders interact with feedback from smart energy monitors in the longer term. Energy Policy **52**(Supplement C), 126–134 (2013). https://doi.org/10.1016/j.enpol.2012.03.027. http://www.sciencedirect.com/science/article/pii/S0301421512002327. Special Section: Transition Pathways to a Low Carbon Economy
15. L. Hattam, D.V. Greetham, Green neighbourhoods in low voltage networks: measuring impact of electric vehicles and photovoltaics on load profiles. J. Mod. Power Sys. Clean Energy **5**(1), 105–116 (2017)
16. L. Hattam, D.V. Greetham, An innovation diffusion model of a local electricity network that is influenced by internal and external factors. Phys. A: Stat. Mech. Appl. **490**(Supplement C), 353–365 (2018)
17. L. Hattam, D.V. Greetham, S. Haben, D. Roberts, Electric vehicles and low voltage grid: impact of uncontrolled demand side response in *24th International Conference and Exhibition on Electricity Distribution (CIRED)*, In press, 2017
18. K. Haynes, V. Mahajan, G. White, Innovation diffusion: a deterministic model of space-time integration with physical analog. Socio-Econ. Plann. Sci. **11**(1), 25–29 (1977)
19. HM Government: Smart meter roll-out (GB): cost-benefit analysis (Nov 2016). https://www.gov.uk/government/publications/smart-meter-roll-out-gb-cost-benefit-analysis. [online], Accessed 20 April 2017
20. J. Hogg, J. Scott, On the use of suboptimal matchings for scaling and ordering sparse symmetric matrices. Numer. Linear. Algebra. Appl. **22**(4), 648–663 (2015). https://doi.org/10.1002/nla.1978. Nla.1978
21. T. Hong, S. Fan, Probabilistic electric load forecasting: a tutorial review. Int. J. Forecasting **32**(3), 914–938 (2016). https://doi.org/10.1016/j.ijforecast.2015.11.011. http://www.sciencedirect.com/science/article/pii/S0169207015001508
22. J. de Hoog, V. Muenzel, D.C. Jayasuriya, T. Alpcan, M. Brazil, D.A. Thomas, I. Mareels, G. Dahlenburg, R. Jegatheesan, The importance of spatial distribution when analysing the impact of electric vehicles on voltage stability in distribution networks. Energy. Sys. **6**(1), 63–84 (2015). https://doi.org/10.1007/s12667-014-0122-8
23. R.J. Hyndman, A.B. Koehler, Another look at measures of forecast accuracy. Int. J. Forecasting **22**(4), 679–688 (2006). https://doi.org/10.1016/j.ijforecast.2006.03.001. http://www.sciencedirect.com/science/article/pii/S0169207006000239
24. E. Karakaya, Finite Element Method for forecasting the diffusion of photovoltaic systems: Why and how? Appl. Energy **163**, 464–475 (2016)

25. E. Kiesling, M. Günther, C. Stummer, L. Wakolbinger, Agent-based simulation of innovation diffusion: a review. Central Eur. J. Oper. Res. **20**(2), 183–230 (2012)
26. R. McKenna, L. Hofmann, E. Merkel, W. Fichtner, N. Strachan, Analysing socioeconomic diversity and scaling effects on residential electricity load profiles in the context of low carbon technology uptake. Energy Policy **97**, 13–26 (2016). https://doi.org/10.1016/j.enpol.2016.06.042. http://www.sciencedirect.com/science/article/pii/S0301421516303469
27. A. Molderink, V. Bakker, M.G.C. Bosman, J.L. Hurink, G.J.M. Smit, A three-step methodology to improve domestic energy efficiency, in *2010 Innovative Smart Grid Technologies (ISGT)*, 1–8 2010. https://doi.org/10.1109/ISGT.2010.5434731
28. R. Morrill, The shape of diffusion in space and time. Econ. Geography **46**, 259–268 (1970)
29. J. Munkres, Algorithms for the assignment and transportation problems. J. Soc. Ind. Appl. Math. **5**, 32–38 (1957)
30. M. Neaimeh, R. Wardle, A.M. Jenkins, J. Yi, G. Hill, P.F. Lyons, Y. Hbner, P.T. Blythe, P.C. Taylor, A probabilistic approach to combining smart meter and electric vehicle charging data to investigate distribution network impacts. Appl. Energy **157**, 688–698 (2015). https://doi.org/10.1016/j.apenergy.2015.01.144. http://www.sciencedirect.com/science/article/pii/S0306261915001944
31. Pecan Street Inc: Dataport Pecan street (Nov 2016). http://dataport.pecanstreet.org. [online], accessed 20 April 2017
32. S. Quoilin, K. Kavvadias, A. Mercier, I. Pappone, A. Zucker, Quantifying self-consumption linked to solar home battery systems: Statistical analysis and economic assessment. Appl. Energy **182**, 58–67 (2016). https://doi.org/10.1016/j.apenergy.2016.08.077. http://www.sciencedirect.com/science/article/pii/S0306261916311643
33. E. Rogers, *Diffusion of Innovations* (The Free Press, New York, 1962)
34. M. Rowe, T. Yunusov, S. Haben, C. Singleton, W. Holderbaum, B. Potter A peak reduction scheduling algorithm for storage devices on the low voltage network. IEEE Trans. Smart Grid **5**(4), 2115–2124 (2014). https://doi.org/10.1109/TSG.2014.2323115
35. M. Sathe, O. Schenk, H. Burkhart, An auction-based weighted matching implementation on massively parallel architectures. Parallel Comput. **38**(12), 595–614 (2012). https://doi.org/10.1016/j.parco.2012.09.001. http://www.sciencedirect.com/science/article/pii/S0167819112000750
36. A. Schrijver, *Combinatorial Optimization: Polyhedra and Efficiency* (Springer, Berlin Heidelberg, 2002)
37. M. Sharifzadeh, H. Lubiano-Walochik, N. Shah, Integrated renewable electricity generation considering uncertainties: The UK roadmap to 50% power generation from wind and solar energies. Renewable Sustainable Energy Rev. **72**, 385–398 (2017). https://doi.org/10.1016/j.rser.2017.01.069. http://www.sciencedirect.com/science/article/pii/S1364032117300795
38. K. Shinohara, H. Okuda, Dynamic innovation diffusion modelling. Comput. Econom. **35**(1), 51 (2009)
39. R.P. Singh, P.X. Gao, D.J. Lizotte, On hourly home peak load prediction, in *2012 IEEE Third International Conference on Smart Grid Communications (SmartGridComm)*, 163–168 2012. https://doi.org/10.1109/SmartGridComm.2012.6485977
40. B.K. Sovacool, P. Kivimaa, S. Hielscher, K. Jenkins, Vulnerability and resistance in the united kingdom's smart meter transition. Energy Policy **109**(Supplement C), 767–781 (2017). https://doi.org/10.1016/j.enpol.2017.07.037. http://www.sciencedirect.com/science/article/pii/S0301421517304688
41. M. Sugiyama, Climate change mitigation and electrification. Energy Policy **44**, 464–468 (2012). https://doi.org/10.1016/j.enpol.2012.01.028. http://www.sciencedirect.com/science/article/pii/S030142151200033X
42. C. Sun, F. Sun, S.J. Moura, Nonlinear predictive energy management of residential buildings with photovoltaics and batteries. J. Power Sources **325**, 723–731 (2016). https://doi.org/10.1016/j.jpowsour.2016.06.076. http://www.sciencedirect.com/science/article/pii/S0378775316307789

43. C. Swinerd, K. McNaught, Design classes for hybrid simulations involving agent-based and system dynamics models. Simul. Modell. Practice Theory **25**(Supplement C), 118–133 (2012)
44. C. Swinerd, K. McNaught, Simulating the diffusion of technological innovation with an integrated hybrid agent-based system dynamics model. J. Simul. **8**(3), 231–240 (2014)
45. J.W. Taylor, A. Espasa, Energy forecasting. Int. J. Forecasting **24**, 561–565 (2008)
46. Taylor, J.W., de Menezes, L.M., McSharry, P.E.: A comparison of univariate methods for forecasting electricity demand up to a day ahead. Int. J. Forecasting **22**(1), 1–16 (2006). https://doi.org/10.1016/j.ijforecast.2005.06.006. http://www.sciencedirect.com/science/article/pii/S0169207005000907
47. N. Tomizawa, On some techniques useful for solution of transportation network problems. Networks **1**, 33–34 (1971)
48. Tzafestas, S., Tzafestas, E.: Computational intelligence techniques for short-term electric load forecasting. J. Intell. Rob. Sys. **31**(1), 7–68 (2001). https://doi.org/10.1023/A:1012402930055
49. J.D. Watson, N.R. Watson, D. Santos-Martin, A.R. Wood, S. Lemon, A.J.V. Miller, Impact of solar photovoltaics on the low-voltage distribution network in New Zealand. IET Generation, Transmission Distribution **10**(1), 1–9 (2016). https://doi.org/10.1049/iet-gtd.2014.1076

Danica Vukadinović Greetham is Lecturer in Mathematics and Director of the Centre for the Mathematics of Human Behaviour. Her main work is in developing methods for modelling, analysis and prediction of human behaviour using large datasets such as social media, smart meters, club cards etc. Her research interests include network analysis, algorithmic graph theory, and forecasting individual electric demand.

Laura Hattam is a Postdoctoral Researcher within the Centre for the Mathematics of Human Behaviour at the University of Reading. Her research interests include dynamical systems theory and its applications, and mathematical modelling.

Chapter 8
Big Data Improves Visitor Experience at Local, State, and National Parks—Natural Language Processing Applied to Customer Feedback

Hari Prasad Udyapuram and Srinagesh Gavirneni

Abstract Local, State and National parks are a major source of natural beauty, fresh air, and calming environs that are being used more and more by visitors to achieve mental and physical wellbeing. Given the popularity of social networks and availability of smartphones with user-friendly apps, these patrons are recording their visit experiences in the form of online reviews and blogs. The availability of this voluminous data provides an excellent opportunity for facility management to improve their service operation by cherishing the positive compliments and identifying and addressing the inherent concerns. This data however, lacks structure, is voluminous and is not easily amenable to manual analysis necessitating the use of Big Data approaches. We designed, developed, and implemented software systems that can download, organize, and analyze the text from these online reviews, analyze them using Natural Language Processing algorithms to perform sentiment analysis and topic modeling and provide facility managers actionable insights to improve visitor experience.

8.1 Introduction

> There is a pleasure in the pathless woods,
> There is a rapture on the lonely shore,
> There is society, where none intrudes,
> By the deep Sea, and music in its roar:
> I love not Man the less, but Nature more,
>
> —*Lord Byron, "Childe Harold's Pilgrimage"*.

H. P. Udyapuram · S. Gavirneni (✉)
Samuel Curtis Johnson Graduate School of Management, Cornell University, Ithaca, NY 14853, USA
e-mail: nagesh@cornell.edu

H. P. Udyapuram
e-mail: hari.uday@gmail.com

A. Emrouznejad and V. Charles (eds.), *Big Data for the Greater Good*,
Studies in Big Data 42, https://doi.org/10.1007/978-3-319-93061-9_8

As the pace of day-to-day life moves faster, the living and work environment gets more mechanized and electronic, the food and the clothing become artificial, the desire to re-connect with nature becomes stronger. This has manifested into a multi-billion dollar global nature-based tourism that continues to grow [5]. While this industry (a key ingredient of many countries' economic policy) is mostly centered around national, state and local parks and other protected areas which are abound with natural beauty, calming surroundings, and tranquil atmosphere; for this industry to continue to grow and thrive, there is a strong need for using scientific and data-driven [16] service operations management techniques. All around the world, there are more than thirty thousand parks and protected areas covering more than thirteen million square kilometers [8]. This represents about nine percent of total land area of the planet. These establishments can be found in over one hundred countries and their dependent territories. Of course, this growth did not occur overnight and has in fact been building (somewhat exponentially) over the past hundred or so years [3].

When managed well, ecotourism (another name for this industry) has the potential [23] to generate jobs, revenues (some in the form of foreign exchange), philanthropic donations, and better health outcomes. While the designation, development, promotion, and maintenance of national parks and protected areas is a multi-faceted challenge that should involve experts from the travel industry, ecological sciences, environmental policy and safety and security, the quantification, tracking, and improvement of the customer visits is often a second thought [13]. However, this is changing and there is increasing inclination towards customer-centric strategies. Parks and Preserves that are known exclusively for their scenic beauty, bio-diversity and recreational activities are now developing reputations as places that people want to visit because of the way they are received, the tremendous experiences awaiting them, the memories they expect to build, the facilities and processes that are hassle-free, and staff that welcomes them with thoughtful actions. The on-going challenge that such parks face is to be a choice destination to a wide variety of people, while simultaneously upholding the highest standards of service and stewardship. In order to achieve this balance, parks have to think about the needs and interests of the people that frequent them and those they want to serve.

Most locations and their management often have very rudimentary (e.g., comment cards) approaches to collect customer feedback and even for the small number of responses they receive, do not have the adequate resources to analyze them. Given the recent technological developments in communication, data collection and storage, and text analytics, we propose, implement, and demonstrate the effectiveness of a novel customer feedback analysis system.

Park visitors are just like those to any service (hotel, restaurant, hospital, theater, etc.) in the sense that they come with certain initial expectation on facilities, duration, and quality. In order to ensure that these expectations can be met with a high level of guarantee, it is important for the service provider (in this case the park management) to not only know these expectations, but actively manage them and setup an environment that consistently beats them. The degree to which these expectations are met and hopefully exceeded determines whether the customers re-visit, tell their relatives, friends and acquaintances to visit, and volunteer their time and resources.

Receiving customer feedback, measuring customer satisfaction, and identifying and fixing inherent issues has been a salient issue in a wide array of industries such as hotels [1, 12, 21], movies [4], and others [10, 17], for the past few decades and it is time for the same management philosophy to be brought over and applied to park management.

There are many dimensions in which the customer experience can be measured. The most important being (i) visual attractiveness and appeal of the scenic areas; (ii) condition of the physical facilities (such as bathrooms, fences, trails); (iii) congestion levels; (iv) staff appearance and demeanor; (v) promptness and accuracy of information; (vi) safety and security; (vii) availability of amenities (restaurant, café, drinks, etc.); (viii) pricing and value for money; (ix) entertainment and recreation; and (x) friendliness to families and children.

Effective management of parks and protected areas necessitates gathering the information on where visitors are going, what they are doing, who they are, where they are coming from, how satisfied they were with the visit, what they liked most, what they liked the least, and what suggestions they may have for improvement. Comprehensive data collection efforts followed by rigorous analysis can ensure an accurate assessment of the values of the park, its resources and its activities. The techniques used for data collection must clearly support the underlying objectives and traditionally many organizations have resorted [7] to physical questionnaires, telephone and online surveys, structured interviews, and paid and unpaid focus groups. Visitor surveys are often seen as a lower-cost, higher effective approach to collecting customer feedback. They are so popular that, now park agencies worldwide are using them and this has resulted in the ability for the agencies to benchmark themselves against one another. In addition, this also enabled the park agencies to communicate their accomplishments to stakeholders (e.g. tax payers, policy makers), accurately and comprehensively. It is well known that many such surveys conducted by park management have had low responses (often in single digit percentages) and those that were received showed significant bias [6]. Usually only those visitors that have had a strongly negative experience tended to respond. There is clearly a need to develop a systematic approach to collecting and analyzing large volumes of accurate customer feedback. One interesting and notable trend is that visitors are increasingly using social networking sites to register their experience in a variety of forms such as online reviews, posts, tweets, images, videos, hashtags. This digital content which is growing at a striking pace not only influences other potential visitors but also contains useful insights for park management to tap into.

Having evolved from the early systems such as newsgroups, listservs, chat rooms, messenger services, social networking has undoubtedly changed the way we communicate, and reduced the cost and inhibition associated with sharing feelings, experiences, and activities. The advent of the smartphone, the meteoric rise in its adoption, the availability of low/no cost apps that enable easy sharing of comments, photos, and hashtags has further encouraged and enabled information sharing. It is rather astonishing to see the large number of people sharing vast amounts of data (which no longer necessarily implies numbers, but could be in various formats such as text, images, weblinks, hashtags) on these social networks. Park management could and should utilize this data for strengthening their programs, formulating marketing and outreach strategies and improving services. Big Data presents serious challenges around data collection, storage and integration. Most of the data is noisy and unstructured. In order to fully harness the power of Big Social Data, parks management have to build proper mechanisms and infrastructure to deal with data.

In this chapter, we demonstrate how machine learning based natural language processing methods can be used to analyze social media content of park visitor comments. We do so by using the New York State Park System in the United States of America as an example. We describe the software methodologies and infrastructure used to extract useful and actionable insights for park management, economic development and service operations.

8.2 New York State Park System

The New York State (NYS) park system consists of 214 parks and historic sites, over 2000 miles of trails, 67 beaches, and 8355 campsites. It attracts approximately 60 million visitors every year. The State Office of Parks, Recreation, and Historic Preservation is responsible for operating and maintaining the state park system, and one of its strategic priorities is to "Increase, Deepen, and Improve the Visitor Experience". Visitor feedback is integral to achieving this objective, but traditional feedback methods—public meetings, web-based surveys and comment cards—are often tedious, expensive, and limited by low participation. Public online review platforms such as TripAdvisor offer a large volume of visitor feedback that could vastly improve how NYS park managers and other community leaders concerned with tourism or business development currently understand and improve visitor experiences.

8.3 The Challenges of Utilizing Public Online Reviews

The NYS park system could develop a deeper understanding of diverse public opinions about its parks by harnessing public online reviews. However, the data on these sites lacks structure, is voluminous, and is not easily amenable to manual analysis. In order to tap into this rich source of visitor information, facility managers need new ways to get feedback online and new software tools for analysis.

A research group from Cornell's Samuel Curtis Johnson Graduate School of Management and the Water Resources Institute designed, developed, and implemented software systems [22] to tap online review content for the benefit of state agencies and the general public. Among the many online social platforms that host visitor reviews, Yelp, TripAdvisor, and Google are the most popular, and thus were used to develop a pilot decision support system for NYS park managers.

8.4 Basics of Natural Language Processing

We briefly describe some of the popular analytical techniques associated with natural language processing and list the various available software systems [9, 19] that can utilized to perform the proposed analysis.

8.4.1 Preprocessing and Text Representation

Arguably the first critical step in analyzing unstructured and voluminous text is the transformation of the free form (qualitative) text into a structured (quantitative) form that is easier to analyze. The easiest and most common transformation is with the "bag of words" representation of text. The moniker "bag of words" is used to signify that the distribution of words within each document is sufficient, i.e., linguistic features like word order, grammar, and other attributes within written text can be ignored (without losing much information) for statistical analysis. This approach converts the corpus into a *document-term matrix* [11]. This matrix contains a column for each word and a row for every document and a matrix entry is a count of how often the word appears in each document.

The resulting structured and quantitative document-term matrix can then, in principle, be analyzed using any of the available mathematical techniques. The size of the matrix, however, can create computational and algorithmic challenges. Natural language processing overcomes this hurdle by emphasizing meaningful words by removing uninformative ones and by keeping the number of unique terms that appear in the corpus from becoming extremely large. There are preprocessing steps that are standard, including (i) transforming all text into lowercase, (ii) removing words composed of less than 3 characters and very common words called stop words (e.g., the, and, of), (iii) stemming words, which refers to the process of removing suffixes, so that words like values, valued and valuing are all replaced with valu, and finally (iv) removing words that occur either too frequently or very rarely.

8.4.2 Sentiment Analysis

Word counts and sentiment represent the most basic statistics for summarizing a corpus, and research has shown that they are associated with customer decision making and product sales [10].

To accommodate potential non-linearities in the impact of sentiment on customer behavior, it is recommended to separately estimate measures of positive sentiment and negative sentiment of each item of customer feedback. The positive sentiment score is calculated by counting the number of unique words in the review that matched a list of "positive" words in validated databases (called *dictionaries*) in the existing literature. The negative sentiment score is calculated analogously.

The choice of dictionary is an important methodological consideration when measuring sentiment, since certain words can have sentiment that changes with the underlying context. For instance, [15] show that dictionaries created using financial 10-K disclosures are more appropriate for financial sentiment analysis rather than dictionaries created from other domains. We chose the dictionaries in [14, 18], since they were created, respectively, to summarize the opinions within online customer reviews and to perform tone analysis of social media blogs. In total, the combined dictionaries consist of approximately 10,000 labeled words.

8.4.3 Probabilistic Topic Modeling

Probabilistic Topic modeling refers to a suite of algorithms that automatically summarize large archives of text by identifying hidden "topics" or themes that are present within [2].

Latent Dirichlet Allocation (LDA) is a popular topic modeling algorithm where the hidden topic structure is inferred using a probabilistic framework. The main idea is that all documents share the same topic set, but each document exhibits a different probabilistic mixture of the said topics. A topic is a collection words and it is presumed that the words are weighted differently in different topics. LDA uses a Bayesian estimation framework to the given text data to infer the topics (distributions over words) and decompose each document into a mixture of topics. Thus, the outputs of LDA are two probability distributions: $P(topic|review)$, the probability distribution of topics in a given review, and $P(word|topic)$, the probability distribution of words within a topic.

From these probability distributions, we estimate two sets of variables for each review i. First, for each topic k, we can estimate a binary variable $Topic_{ik}$, with $Topic_{ik} = 1$ implying that review i discusses topic k. We set $Topic_{ik}$ equal to 1 if the posterior probability $P(k|i)$ is greater than the median probability across all reviews and topics threshold, and 0 otherwise. Second, we computed a continuous variable called "focus", which we define as $Focus_i = \sum_{k=1}^{K} P(k|i)^2$. Reviews with larger values of focus (sometimes as the Herfindahl Index [20]), intuitively speaking, are those that are concentrated on a small number of topics.

8.5 Bear Mountain State Park—A Case Study

To demonstrate the utility of these tools and analytic approaches, an analysis was performed on the reviews collected for Bear Mountain State Park. Located on the west bank of the Hudson River, Bear Mountain attracts more than two million visitors per year and is visited for a wide variety of activities such as fishing, swimming, hiking, and boating. As of June 2014, there were 70 reviews on Yelp, 191 on TripAdvisor, and 34 on Google for this park. The reviews were downloaded into a database that could be searched and sorted in a cursory manual review. Automated analyses in the areas of sentiment rating and topic modeling were then applied to the more than 25,000 words of text found in the data.

Figure 8.1 illustrates the distribution of sentiment among the reviews. About 45% of the reviews have a sentiment rating of 0.7 or higher, mostly positive. As an example, a review that received a sentiment rating of 0.9 contained phrases such as "*beautiful attraction*," "*absolutely gorgeous*," and "*fun to watch*." A review that received a sentiment rating of 0.4 contained phrases such as "*incredibly crowded and dirty*," "*trash all over*," and "*bathrooms were disgusting*."

The negative and positive reviews were then separated and probabilistic topic modeling was performed on both sets of reviews. From the negative reviews, the following major topics (with quotes from related reviews, edited for grammar) were distilled:

Crowded
…The park grounds were incredibly crowded and dirty…
…Arriving at the pool… we had to make a line because it was crowded going in…
Parking
…Worn out, run down, poor access/parking, filled with drunk young punks… parking was $8! I have been to a bunch of parks and this is the first one I've had to pay.
Trash
…There was trash all over the place…
…There were tons of plastic cups and trash (p.s. trash bin literally 3 ft away) on the floor and even in the pond…

From the positive reviews, the following topics were distilled:

Trails
…Great outdoors for all to enjoy all seasons. Trails, zoo and lake walk to be enjoyed…
…There are lots more trails and good bird watching at Iona Island…
Animals

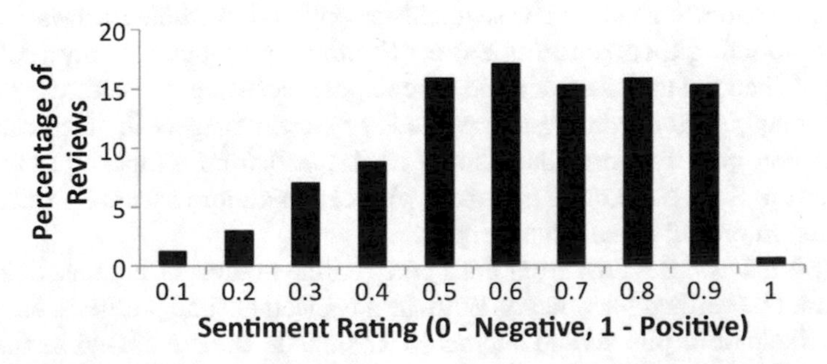

Fig. 8.1 Distribution of sentiment found in the online reviews of New York State Parks

…Took my granddaughter and was thrilled to see all the natural beauty… with its natural rock, flora and animals…
…We were able to enjoy the zoo (which is small, but cute) with plenty of animals…
Hike
…It was a nice hike (LOTS of steps, our butts felt it later!) with a beautiful view at the top…
…Lake, carousel, hikes, play fields, zoo, WOW!…

This type of analysis helps a facility manager to not only identify the overall sentiment about their park and facilities, but to recognize which features should be appropriately leveraged in marketing and future investments. Likewise, the analysis identifies areas that need to be addressed to improve visitor experience. Throughout this process, the topics, themes, and even language used by visitors in reviews can also be used by facility managers to target and improve marketing and communications efforts.

8.6 Review Analysis—Further Enhancements

These analytical tools are continuously improving, algorithmically getting faster, more sophisticated feature and thus are applicable not only to government operated facilities, but also to resources managed by non-profit groups such as land trusts. Depending on the needs of state or local agencies and other managers of public spaces, these tools could be enhanced in the following ways:

Automatic Downloading of Reviews: Online review platforms continue to increase in popularity and new reviews are submitted on a regular basis. Manual downloading is time consuming, especially for managers who want "real time" reports. It is necessary to develop and implement a process that automatically downloads and organizes online reviews into a database.

Topic Modeling on Negative/Positive Review Segments: The current approaches extracts themes from whole reviews that have been labeled as positive or negative. But reviews are rarely, if ever, completely positive or negative. Each review typically contains segments that are positive alongside segments that are negative. In order to get a more accurate collection of themes, the analysis should perform topic modeling on collections of review segments as opposed to whole reviews.

Topic Modeling Incorporating Expert Feedback: A topic is simply a collection of words. When the topics are chosen by computer software, some of the words in the topic may not fit according to the needs of park managers. In such cases, the managers can identify words that should be dropped from a topic and the model can be re-run. Such a recursive approach will lead to a more accurate extraction of themes and improved managerial insights.

Verify Reviews: Reviews from third party online platforms, are unsolicited and often cannot be verified for veracity. With the advancement and proliferation of technologies like mobile phones and microchip wristbands, the use of devices that track key personal information are increasingly common. These devices carry important information which visitors could voluntarily share with the facility management to create verified or more detailed reviews.

Identifying and Accommodating Temporal Changes: Instances exist in which the underlying data of reviewer characteristics, the length and content of reviews, the topics discussed and even the language used can undergo a seismic shift. When and if that happens, the existing analysis, applied without any changes, can lead to wrong insights and conclusions. It is necessary to have an approach for identifying such temporal shifts in data and determine ways in which the analysis should be appropriately adjusted.

8.7 Conclusion

The necessity and usefulness of natural scenic areas such as local, state, and national parks, has never been greater for our society. They provide much needed serenity and tranquility and resulting mental health benefits and stress relief in our lives that are increasingly getting lonely, technology-centric, and hectic. Given that these benefits should be realized in short durations of available free time, it is important that appropriate information and feedback is provided to the visitors. The traditional methods with which the park managers connect with their visitors have had many strengths, but also numerous weaknesses, and the availability of social networks and the information available within them can further strengthen the communication between the park visitors and park management. Given the new developments in data storage, computational capacity, and efficiency of natural language processing algorithms, a new opportunity is presented to the park managers to better define visitor experience, measure their capability to meet these expectations, and make the necessary changes to improve their service offering. Using the New York State park system as the backdrop, we demonstrate the system we built to apply natural language processing algorithms to customer feedback collected from the various social networks. This is only a small first step and we describe ways in which such systems can be further improved and implemented by the teams managing these local, state, and national parks.

Acknowledgements This research was conducted at the New York State Water Resources Institute and funded by the Hudson River Estuary Program, a program of the NYS Department of Environmental Conservation. We are grateful for the support we received from those two organizations on the intellectual, educational, technological, motivational, and financial dimensions.

References

1. I. Blal, M.C. Sturman, The differential effects of the quality and quantity of online reviews on hotel room sales. Cornell Hosp. Q. **55**(4), 365–375 (2014)
2. D. Blei, J. Lafferty J, Topic models, in *Text Mining: Classification, Clustering, and Applications*, ed, by A. Srivastava, M. Sahami (Chapman & Hall/CRC Data Mining and Knowledge Discovery Series, 2009), pp. 71–94
3. S. Driml, M. Common, Economic and financial benefits of tourism in major protected areas. Aust. J. Enviro. Manage. **2**(2), 19–39 (1995)
4. W. Duan, B. Gu, A.B. Whinston, The dynamics of online word-of-mouth and product sales—An empirical investigation of the movie industry. J. Retail. **84**(2), 233–242 (2008)
5. P.F.J. Eagles, Trends in park tourism: economics, finance, and management. J. Sustain. Tourism **10**(2), 132–153 (2002)
6. C.M. Fleming, M. Bowden, Web-based surveys as an alternative to traditional mail methods. J. Environ. Manage. **90**(4), 284–292 (2009)
7. A.R. Graefe, J.D. Absher, R.C. Burns, Monitoring visitor satisfaction: a comparison of comment cards and more in-depth surveys, in *Proceedings of the 2000 Northeastern Recreation Research Symposium* (USDA Forest Service General Technical Report NE-276, 2000), pp. 265–69
8. M.J.B. Green, J. Paine, State of the world's protected areas at the end of the twentieth century. Paper presented at protected areas in the 21st century: from islands to networks. World Commission on Protected Areas, Albany, WA, Australia (1997)
9. B. Gruen, K. Hornik, Topic models: an R package for fitting topic models. J. Stat. Softw. **40**(13), 1–30 (2011). http://www.jstatsoft.org/v40/i13/
10. N. Hu, N.S. Koh, S.K. Reddy, Ratings lead you to the product, reviews help you clinch it? The mediating role of online review sentiments on product sales. Decis. Support Syst. **57**, 42–53 (2014)
11. R. Kosala, H. Blockeel, Web mining research: a survey. ACM SIGKDD Explor. Newslett. **2**(1), 1–15 (2000)
12. S.E. Levy, W. Duan, S. Boo, An analysis of one-star online reviews and responses in the Washington, D.C., lodging market. Cornell Hosp. Q. **54**(1), 49–63 (2013)
13. K. Lindberg, Economic aspects of ecotourism, in *Ecotourism: A Guide for Planners and Managers*, vol. 2., ed. by M. Epler Wood, K. Lindberg (Ecotourism Society, North Bennington, VT, 1998)
14. B. Liu, Sentiment analysis and subjectivity, in *Handbook of Natural Language Processing*, 2nd edn., ed. by N. Indurkhya, F.J. Damerau (Chapman & Hall/CRC Machine Learning & Pattern Recognition, 2010), pp. 627–666
15. T. Loughran, B. McDonald, When is a liability not a liability? Textual analysis, dictionaries, and 10-ks. J. Finan. **66**(1), 35–65 (2011)
16. S. Mankad, H. Han, J. Goh, S. Gavirneni, Understanding online hotel reviews through automated text analysis. Serv. Sci. **8**(2), 124–138 (2016)
17. S.M. Mudambi, D. Schuff, What makes a helpful online review? A study of customer reviews on Amazon.com. MIS Q. **34**(1), 185–200 (2010)
18. F.A. Nielsen, A new ANEW: evaluation of a word list for sentiment analysis in microblogs, in *Proceedings of the ESCW2011 Workshop on 'Making Sense of Microposts': Big Things Come in Small Packages* (2011), pp. 93–98
19. R Core Team, R: a language and environment for statistical computing, in *R Foundation for Statistical Computing* (Vienna, Austria, 2014). http://www.R-project.org/
20. S.A. Rhoades, The Herfindahl-Hirschman index. Fed. Reserve Bull. March, 188–189
21. B.A. Sparks, V. Browning, The impact of online reviews on hotel booking intentions and perception of trust. Tourism Manage. **32**(6), 1310–1323 (2011)
22. H.P. Udyapuram, S. Gavirneni, Automated analysis of online reviews to improve visitor experience in New York State Parks. *WRI Final Report* (2014). https://wri.cals.cornell.edu/sites/wri.cals.cornell.edu/files/shared/documents/2014_Gavirneni_Final.pdf

23. M.P. Wells, *Economic Perspectives on Nature Tourism, Conservation and Development* (Environment Department Paper No. 55, Pollution and Environmental Economics Division, World Bank, Washington, DC, 1997)

Hari Prasad Udyapuram is an independent researcher and was a visiting research scholar (2013–2016) at the Samuel Curtis Johnson Graduate School of Management at Cornell University. His interest lies in Big Data Analytics, optimization and building algorithms for large-scale online platforms. He is an Information technology consultant for Indian Railways and has held visiting positions at various universities. He has a bachelor's degree in computer science from Indian Institute of Technology, Bombay. He can be contacted at hari.uday@gmail.com.

Nagesh Gavirneni a professor of operations management in the Samuel Curtis Johnson Graduate School of Management at Cornell University. His research interests are in the areas of supply chain management, inventory control, production scheduling, simulation and optimization. He is now using these models and methodologies to solve problems in healthcare, agriculture and humanitarian logistics in developing countries. Previously, he was an assistant professor in the Kelley School of Business at Indiana University, the chief algorithm design engineer of SmartOps, a Software Architect at Maxager Technology, Inc. and a research scientist with Schlumberger. He has an undergraduate degree in Mechanical Engineering from IIT-Madras, a Master's degree from Iowa State University, and a Ph.D. from Carnegie Mellon University.

Chapter 9
Big Data and Sensitive Data

Kurt Nielsen

Abstract Big Data provides a tremendous amount of detailed data for improved decision making, from overall strategic decisions, to automated operational micro-decisions. Directly, or with the right analytical methods, these data may reveal private information such as preferences and choices, as well as bargaining positions. Therefore, these data may be both personal or of strategic importance to companies, which may distort the value of Big Data. Consequently, privacy-preserving use of such data has been a long-standing challenge, but today this can be effectively addressed by modern cryptography. One class of solutions makes data itself anonymous, although this degrades the value of the data. Another class allows confidential use of the actual data by Computation on Encrypted Data (CoED). This chapter describes how CoED can be used for privacy-preserving statistics and how it may distort existing trustee institutions and foster new types of data collaborations and business models. The chapter provides an introduction to CoED, and presents CoED applications for collaborative statistics when applied to financial risk assessment in banks and directly to the banks' customers. Another application shows how MPC can be used to gather high quality data from, for example,. national statistics into online services without compromising confidentiality.

9.1 Introduction

Big Data is large amounts of detailed data that makes it possible to improve data-driven decisions, from overall strategic decisions to automated operational micro-decisions.

The idea of data-driven decision making is neither unique or new. Within areas like financial risk assessments and fraud detection and logistics, data-driven decision

---

K. Nielsen (✉)
Department of Food and Resource Economics,
University of Copenhagen, Rolighedsvej 25, Frederiksberg C 1958, Denmark
e-mail: kun@ifro.ku.dk

K. Nielsen
Partisia, Aabogade 15, Aarhus N 8200, Denmark
e-mail: kn@partisia.com

A. Emrouznejad and V. Charles (eds.), *Big Data for the Greater Good*,
Studies in Big Data 42, https://doi.org/10.1007/978-3-319-93061-9_9

making has been best practice for decades. The new perspectives are the increasing amount of data that comes from the digitalization of almost everything. These more or less endless data are revealing increasingly greater detail about almost all aspects of the economy and our lives - even our genomes.

These data can help us in making better decisions that can improve our lives and improve the utilisation of our scarce resources. From an economics perspective, all decisions can be boiled down to a choice from a set of alternatives. With Big Data, we basically get more detailed information about the alternatives, as well as the preferences regarding these alternatives. This makes it possible to guide individuals to make better decisions and to better predict future outcomes.

Sensitive personal and company information is highly valued in research and services, with benefits for individual citizens, companies and society in general. However, the most valuable data are also the most sensitive, such as information about individuals' and companies' private preferences and decisions. On the one hand, it is predicted that data-driven decisions and analytics will be a tremendous growth area in the years to come. On the other hand, data which are used outside their original context may violate fundamental rights to privacy and weaken the "bargaining position" of individuals and companies in general.

The latter was acknowledged early on by Google's chief economist, Hal Varian, in an early paper on market design for automated trading: "*... Hence privacy appears to be a critical problem for computerized purchasing agents. This consideration usually does not arise with purely human participants, since it is generally thought that they can keep their private values secret. Even if current information can be safeguarded, records of past behaviour can be extremely valuable, since historical data can be used to estimate willingness to pay. What should be the technological and social safeguards to deal with this problem?*" [1].

Increasing political awareness has resulted in new regulation that is primarily aimed at protecting personal data. The most progressive example is the General Data Protection Regulation (GDPR) in the EU, which will come into effect in May 2018. The GDPR lists a number of requirements on how to use so-called "Personal Identifiable Information", and introduces penalties for data breaches that align data protection with anti-trust regulation. Data protection outside the EU is also developing in the same direction in response to increasing concerns from citizens and other political pressures. This type of regulation impacts many companies as personal information is an integrated part of their businesses. Sensitive company information is, on the other hand, not regulated in the same way as personal identifiable information. Indirectly, anti-trust regulation prevents sensitive data from being shared among competitors, which may otherwise hamper competition. Figure 9.1 provides a simple overview of sensitive data regulation, as well as incentives for individuals or companies to engage in collaborations that involve sharing sensitive data.

While data protection regulation such as the GDPR makes some existing uses of Big Data illegal, the large potential gains from utilizing Big Data remain. The solution is to not stop using Big Data, but to find a third way that allows its positive use. As a result, privacy-preserving use of such data has been a long-standing challenge.

Fig. 9.1 Types of sensitive data and regulation

	Personal data	Company data
Regulation	Data protection regulation e.g. GDPR (EU)	Anti-trust regulation
Risk assessment	As different as humans	Driven by incentives

Fortunately, today, modern cryptography provides principles, protocols and tools that effectively address this challenge.

One class of solutions makes data itself anonymous, although this degrades the value of the data. Another class allows an analysis of the sensitive data in great detail by the use of "Computation on Encrypted Data" (CoED). This chapter describes how CoED can be used for privacy-preserving statistics, and how it may distort the way data are shared and used today.

The chapter is organized as follows: Sect. 9.2 introduces CoED and how and where it is used. Section 9.3 presents three use cases of CoED within privacy-preserving statistics and Sect. 9.4 discusses the managerial implications of the use of CoED. Section 9.5 discusses and concludes the chapter.

9.2 Compute on Encrypted Data

This section describes the use of one of the most fundamental ways of solving the problem of how to utilize sensitive data, which is simply to keep data encrypted at all times. CoED covers a number of different encryption techniques. Within hardware security, solutions like INTEL's SGX processors allow encapsulated computation within a CPU. However, this solution has a number of known weaknesses,[1] but most importantly, trust is placed on the hardware producers, and hardware is also less transparent and verifiable than a software code. Within software, the most efficient class of solution is, by far, Secure Multiparty Computation (MPC), which keeps sensitive data encrypted at all times without leaving the key to unlock the data with a single person or organisation.[2]

MPC basically allows us to construct the ideal "Trusted Third Party" (TTP) that confidentially coordinates private information according to a comprehensive

[1]Known weaknesses are physical side-channels and information leakages through memory access patterns, which breaks the security. A sound review of SGX is provided by Costan and Devadas [2].
[2]Gartner has MPC (Multiparty Computation) listed as a technology "on the rise" in a recent report on data security in July 2017 https://www.gartner.com/doc/3772083/hype-cycle-data-security.

protocol. The notion of a TTP has been a common construct in information economics since the original work on the Revelation Principle (see, e.g., [3] or [4]). Based on the revelation principle, any equilibrium outcome that is based on any mechanism can also be arranged as the outcome of a direct revelation game in which the participants have an incentive to honestly reveal their private information. However, while economic theory has little to say about the design of such a TTP, it has been a central research question in computer science for decades. Unlike traditional cryptography, which focuses on ensuring privacy within a group of individuals with full access to the information, recent contributions break with the idea of placing all trust in a single entity at any time. MPC belongs to this new generation of cryptographic solutions and allows a number of parties to jointly perform computation on private inputs, without releasing information other than that which has been agreed upon a priori. The seminal aspects of this concept can be traced back to [5], with the theory being founded in the 1980s, see [6]. However, the involved ideas have only recently been refined and made applicable in practice, see [7–9].

9.2.1 Designing MPC Trust

Consider the roles involved in an MPC system; apart from an organizer or an administrator, an MPC system has basically the following 3 roles:

- The Input Parties (**IP**) delivering sensitive data to the confidential computation.
- The Result Parties (**RP**) receiving results or partial results from the confidential computation.
- The Computing Parties (**CP**) jointly computing the confidential computation.

The **IP**s are the data providers who submit encrypted sensitive data to the **CP**s. The **CP**s ensure that the data stays encrypted at all times. The **RP**s are the users of the application that utilize the encrypted data, and receive the results from the approved confidential computations, or the approved application.

The core MPC system consists of n **CP**s as illustrated in Fig. 9.2. There are a number of different MPC protocols and properties within the MPC systems; the list below provides the most important[3]:

- The number of **CP**s involved in the MPC system (n).
- The threshold of **CP**s that cannot run the intended protocol or take control of the MPC system (t), meaning that $t + 1$ **CP**s can use or misuse the system.
- Passive security does not guarantee that a corrupt **CP** computes the intended protocol. With passive security, the **CP**s that make up the MPC system should be mutually trusted.

[3]Other important characteristics are the different types of basic operations such as arithmetic or Boolean, and different types of cryptographic technologies such as secret sharing and homomorphic encryption.

Fig. 9.2 MPC

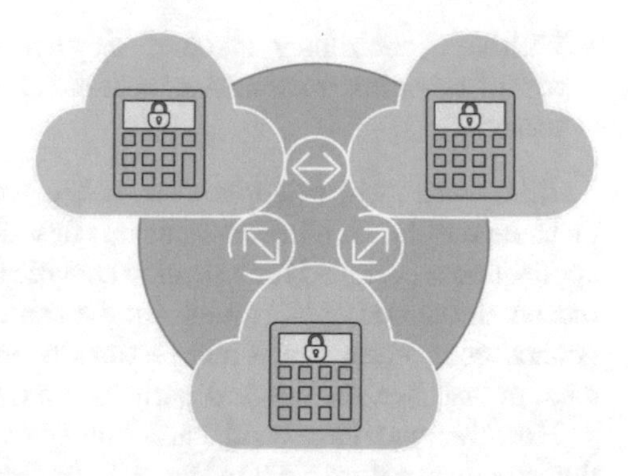

- Active security guarantees that a corrupt **CP** run the intended protocol. With active security, distrusted **CP**s or **CP**s with opposite interests can make up trustworthy MPC system.[4]

The number of **CP**s can be two or more, and the threshold can be any number below n. For $t = n - 1$ it takes all **CP**s to manipulate the MPC system. If the **IP**s are the **CP**s, then the confidentiality cannot be compromised without consensus among all data providers (**IP**s). With active security, the MPC system may consist of **CP**s with opposing interests, such as buyers and sellers, which may be both a cheaper and a stronger way of creating trust as opposed to third parties. The computation speed is heavily dependent on the choice of parameters. While "majority trust" (e.g. $n = 3$ and $t = 1$) with passive security is the fastest,"self trust" (e.g. $n = 2$ and $t = 1$), with active security, is a more secure system. However, in all cases, there is no single point of trust, as opposed to the traditional trustee in the form of a trusted individual or organization.

From an economic point of view, the MPC approach is analogous to paying a consultancy firm to act as a single TTP. However, there are some fundamental differences between the two ways of coordinating confidential information:

- With the MPC approach, no individual or institution gains access to confidential information, unlike when a consultancy firm is used, where confidential information is revealed to a trusted individual, which opens the door to potential human error and/or corruption.
- With MPC, trust is based on the intentions of the **CP**s. Because collusion requires coordination between more than one **CP**, selecting **CP**s with opposing interests strengthens trust. In contrast, when a consultancy firm is used, opposing interests may lead to a scenario that encourages corruption in order to gain a commercial advantage.

[4]Another property is fault tolerance, which allows the MPC system to continue to operate if a **CP** intentionally or unintentionally fails to operate.

- With MPC, coordination is enabled by a piece of software. Therefore, the marginal cost of repeating trusted coordination is lower than that achieved in a traditional scenario.

MPC is an ideal example of so-called security-by-design or privacy-by-design that supports data minimization as it is described in the GDPR regulation. An MPC application is designed for a specific analytical purpose and the designated **RP**s only receive the agreed upon results. On the contrary, if the MPC system were to allow general computations, the intersections between results could reveal the sensitive data. In that case, the whole security setup would provide a false sense of security.

However, when used with care, MPC provides a holistic end-to-end data protection solution as illustrated in Fig. 9.3. The MPC system makes it possible to merge data from distributed data sources and bring them together for specific analytical purposes. Hereby, MPC allows the continued use of highly valuable, sensitive information by services that can benefit individual citizens, companies, and society as a whole. The high level of security and control allows sensitive data to be used without violating fundamental rights to privacy or the "bargaining positions" of individuals and companies.

In order to explore how MPC works, consider the following simplified problem: Adding the two privately held numbers a and b.

Under so-called public-key cryptography, the sender and recipient have separate keys:; one to lock, which is called the public key, and one to unlock, which is called the secret key. The system works much like a padlock; everyone can lock (encrypt), but only the holder of the secret key can unlock (decrypt). Assume that the secret key is c and the encryption function is $E(x) = c^x$. A TTP can, therefore, receive the encrypted information c^a and c^b from the holders of a and b. If we multiply the encrypted numbers, we get c^{a+b}. Therefore, we can calculate the encrypted numbers. Now, in order to understand the final results, we must know c in order to convert c^{a+b} into $a + b$. However, if we know c, we can also calculate the private information a and b based on the encrypted information, and this will not protect the private information. The key c should, therefore, be kept secret or should be separated into pieces so that

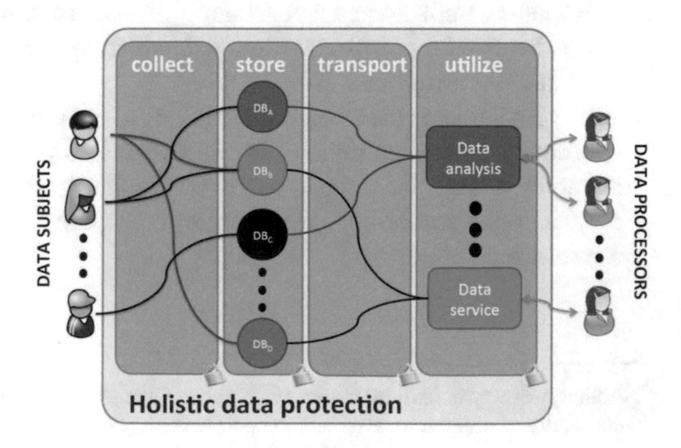

Fig. 9.3 Holistic data protection where data are brought together for specific purposes only

we would need several pieces to actually construct c. This is achieved in MPC by making it necessary for the partial key-holders to collaborate in the calculations.

In order to see how a secret key c can be shared, we can consider it as the solution to a linear function $f(0) = c$. Let $f(x_1) = y_1$ and $f(x_2) = y_2$ be two random function values. Now, by knowing only one of the two numbers, (x_1, y_1) or (x_2, y_2), we have no information about c; however, if we know both, we can determine c. Had $f(.)$ been a second-degree polynomial, any two points would not reveal anything about c, whereas three points would. This approach is known as secret sharing, wherein n is the number of distributed points and t is the largest number of points that does not reveal any information about the private inputs. With n **CP**s, $t + 1$ is the number of TTPs it takes to use (or misuse) the system. In other words, it requires collaboration between at least $t + 1$ of the **CP**s. Figure 9.4 illustrates the estimation of a secret key based on a third-degree polynomial. If each of the four **CP**s, numbered 1,2,3, and 4 on the horizontal axis, gets the private information corresponding to the values on the vertical axis of the points above their numbers, their combined information will be sufficient to determine the secret key, defined as the value on the vertical axis where the polynomial crosses it, i.e. $f(0)$.

A fully functional MPC system can be used for all basic operations, such as addition, subtraction, multiplication, division, and comparisons. While adding numbers is simple, as illustrated above, multiplication, and particularly comparisons, are more complicated. In order to clarify this point, consider the slightly more difficult problem of multiplying two encrypted numbers a and b. Consider an MPC system with $n = 5$ and $t = 2$. As in the example of addition, let the encryption function be $E(x) = c^x$, where c is the distributed secret key, and let each of the five **CP**s generate a random number d_i, $i \in \{1, 2, 3, 4, 5\}$. The d_is are encrypted and broadcasted, so that all TTPs can compute $E(d)$ via simple addition. Also, let all **CP**s compute the encrypted value of $E(a) + E(d)$, and let the results $a + d$ be decrypted. These results reveal nothing because no one knows a or d. Using the number $(a + d)$, $\mathbf{CP}_1$ computes the private number $a_1 = (a + d) - d_1$, and the remaining **CP**s set their private numbers as $a_i = -d_i$, $i \in \{2, 3, 4, 5\}$. Finally, each **CP** computes $E(a_i \cdot b)$ by simply adding $E(b)$, a_i times. Now, adding these five encrypted numbers yields

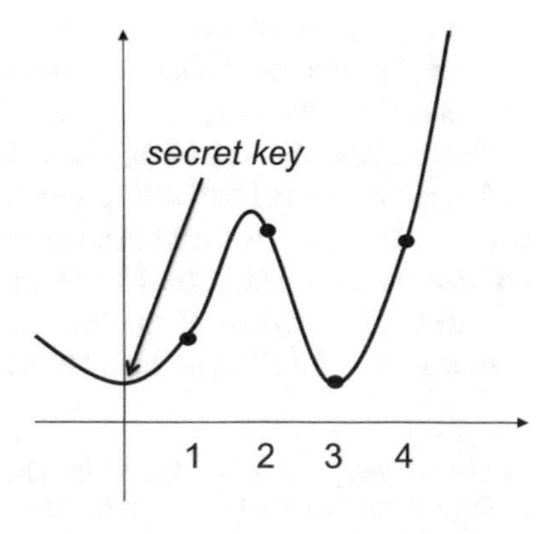

Fig. 9.4 Sharing a secret

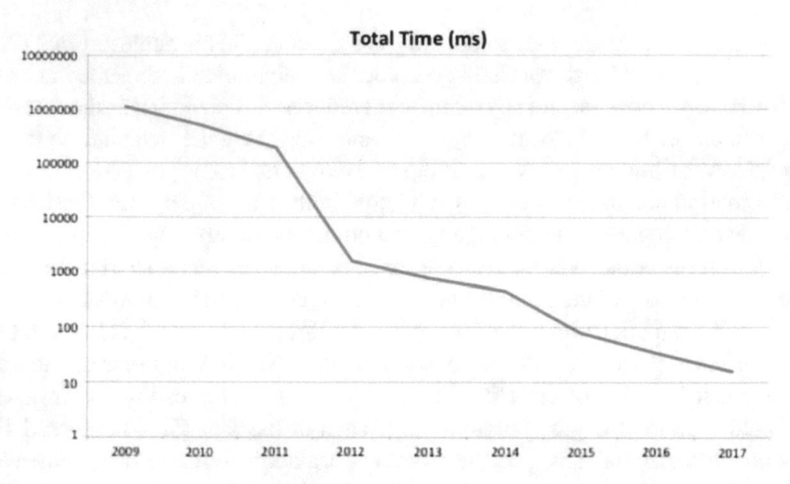

Fig. 9.5 Exponential improved computation time of MPC

the number we are looking for, ab:

$$\prod_{i=1}^{5} c^{a_i b} = c^{a_1 b + a_2 b + a_3 b + a_4 b + a_5 b} = c^{((a+d)-d_1)b - d_2 b - d_3 b - d_4 b - d_5 b} = c^{ab} \tag{9.1}$$

Based on this example, multiplying requires more coordination between the **CP**s. However, the most complicated operations are comparisons as they require significantly more coordination between the **CP**s.

9.2.2 Commercial and Pilot Applications of MPC

Although, MPC was shown, in theory, to be generally applicable in the mid- 1980s, the computational complexity of MPC prevented its practical use for two decades. The first large scale and commercial use of MPC was in 2008 in Denmark, when MPC replaced a traditional auctioneer on a double auction used for the reallocation of production contracts, see e.g. [8, 10].

Since 2008, the technology has matured both in terms of computational speed, and the properties of the MPC protocols. As illustrated in Fig. 9.5, the computational overhead has been reduced exponentially. Figure 9.5 illustrates the extent to which computational time for the same MPC computation has reduced over time.[5]

The highly significant improvements illustrated in Fig. 9.5 have expanded the practical use of MPC, and today a number of companies are investing in it.

[5]The chart was presented by Jesper Buus Nielsen at the opening of the DIGIT center (http://digit.au.dk) and the data points come from the following papers: [11–17].

There is increasing interest and R&D in MPC by a number of major firms. SAP focuses on secure supply chain management. Philips invest in MPC R&D with a focus on healthcare analytics. Google has recently started MPC R&D in privacy-preserving B2B interaction in online advertising. Huawei has also recently invested in MPC, but so far there is no public information about their research. In all cases, the research has focused on prototype software and not commercial products.

The commercial players using MPC are represented by SMEs. Partisia has been the pioneer within commercial MPC services, starting with the aforementioned contract exchange and energy procurement auctions; and recently, off-exchange matching on financial markets for securities and a number of pilot applications within privacy-preserving statistics. In parallel, Cybernetica has developed Sharemind, which is MPC-based statistical software. They has also focused on various pilot applications using large amounts of data, including genome data, satellite trajectories and tax fraud detection. Sepior (a spin-off from Partisia) and Dyadicsec in Israel focus on MPC-based key management and authentication, and have received funding from several private and institutional investors.

Recently, a blockchain company called Enigma has raised funding through a so-called Initial Coin Offering (ICO), which is partly based on promises to combine blockchain and MPC. Another ongoing ICO, KEEP, is currently promising another combination of blockchain and MPC.

In conclusion, the MPC technology offers important properties, and the technology has matured and is ready for privacy-preserving statistics in Big Data scenarios, which is the focus of the next section.

9.3 Privacy-Preserving Statistics

This section describes three prototype and pilot applications of MPC within privacy-preserving statistics. The different applications originate from R&D projects and collaborations between the Danish financial sector and Statistics Denmark.

The financial focus is risk assessment in broad terms; from traditional credit scoring, benchmarking and banks tailored risk assessments, to statistics that help companies self-select into the right financial institutions and products.

9.3.1 Collaborative Statistics for Risk Assessment

Risk assessment is an essential component in any lending institution, both before and after a loan, or other financial product, has been issued. The fundamental problem is the lack of information about the likelihood of an individuals or a companys ability to fulfil its financial obligations. In economics, the two basic problems with this informational asymmetry between the lender and the borrower, are known as adverse selection and moral hazard.

Adverse selection deals with the difficulties in distinguishing between good and bad types before a loan is issued (individuals or firms with either a high or low probability of repaying the loan in full). With no further information, the lender may be forced to offer the same conditions and interest rate to both types, and the higher interest rate may be unattractive to the low risk borrowers, who then find other solutions. Consequently, the high risk borrowers may crowd out the low risk borrowers, and the interest rate may rise even further, ultimately reducing the market significantly. With good data about the different borrowers, the lender may be able to distinguish between low and high risk and offer each a tailor-made contract.

Moral hazard deals with the difficulties in distinguishing between low and high risk borrowers after a loan has been issued. Without more detailed further information, the lender may not be able to predict who will act in a way that increases the risk to the lender (e.g. by low performance and more borrowing, etc). Consequently, the lender may be forced to offer the same conditions to both types of borrower, which again may cause low risk borrowers to find other solutions. Consequently, the high risk borrowers may crowd out the low risk and the interest rate may rise even further etc. However, with good information, the lender may be able to distinguish between low and high risk borrowers and act accordingly to prevent losses.

In both cases, the core of the problem is a lack of information, the solution to which is to find or create trustworthy indicators, e.g. credit scores and benchmarks to be used before a loan is issued, and during the lifetime of a loan. The regulation and the practice around this "production of signals" (e.g. credit scores), differs. In the UK and the US, everyone obtains their own credit score from separate rating firms before entering banks and other financial companies. In contrast, in the rest of the EU, credit scoring is mostly conducted by banks or other financial companies that provide financial products, e.g. loans. In both cases, more information is required to produce better indicators. Recently, less structured and more subjective data have been used in credit ratings. Increasing digital footprints and transaction data are examples of new sources of information about an individuals type or a companies' performance. The use of these sensitive data requires a solution such as MPC.

The following two applications describe how commercial banks deal with risk assessment. In the first case, MPC facilitates the use of existing private information about Danish farmers' accounts and production data. The second case is a more elaborated application for collaborative statistics across banks and the banks' customers.

9.3.1.1 Confidential Benchmarking in Finance

This section presents a prototype application of MPC developed across a number of R&D projects including the Danish research center CFEM[6] and the EU-project PRACTICE[7] and published in [18].

---

[6]http://cs.au.dk/research/centers/cfem.

[7]https://practice-project.eu.

The prototype uses MPC for improved credit rating, while the business case focuses on farmers, which represent a business segment that is particularly challenging for Danish banks. The fundamental problem is that many farmers have had debt ratios that have been too high and earnings that have been too low for years, and the banks have been reluctant to realise the losses.[8] One would like to avoid the emergence of a chain reaction that may affect the individual bank's credit rating and, therefore, its ability to lend money on the interbank market. On the one hand, a number of banks have too many high risk customers. On the other hand, status quo worsens the situation for this group of customers that need to invest to stay competitive. To control the situation and to avoid an escalation, the banks are forced to select the right farmers from this group of risk-prone customers and offer them more loans. This requires more analysis than the traditional credit scores. The banks need to look beyond the debt and identify the farmers who perform better and who are more likely to pay back the loans. To achieve this, one needs accounting and production data from a large range of peer farmers. Since farmers are not required to publish accounting data, a number of banks, in particular the small and medium sized ones, lack the basic data for sound in-house analysis of their customers relative performance.

However, there is a consultancy firm in the market that possesses accounting data from a large number of farmers. The consultancy house is required to keep its database confidential, while the banks are not allowed to divulge their customers data including their identity.

In close collaboration with this consultancy firm and selected small and medium sized banks, MPC was used to bridge the gap and provide a richer data foundation by merging the confidential data from the accounting firm with additional confidential data from the individual banks. An MPC based Linear Programming (LP) solver was used to conduct state-of-the-art relative performance analysis directly on the richer, albeit secret, data set. The resulting benchmarks were used to evaluate new individual customers, as well as the banks' portfolios. In both cases, the analysis reflects performance relative to the agricultural sector as a whole.

In general terms, benchmarking is the process of comparing the performance of one unit against that of best practice. The applied Data Envelopment Analysis (DEA) is a so-called frontier-evaluation technique that supports best practice comparisons (or efficiency analysis) in a realistic multiple-inputs/multiple-outputs framework. DEA was originally proposed by Charnes et al. [19, 20] and has subsequently been refined and applied in a large number of research papers and by consultants in a broader sense. A 2008 bibliography lists more than 4000 DEA references, with more than 1600 of these being published in high quality scientific journals [21].

DEA is usually used to gain a *general insight* into the variation in performance or the productivity development in a given sector. However, DEA has also proven useful in *incentive provision*, for example, regulation and as part of an auction market cf. [22–24]. Finally, DEA has also been applied as a direct alternative to traditional credit rating models in predicting the risk of failures/defaults [25–27]. DEA can

[8]The banks are typically the lenders with the utmost priority in case of default.

be formulated as an LP-problem and can, therefore, be solved by the MPC based LP-solver.

To formally define the DEA methodology used for this analysis, consider the set of n observed farms. All farms are using r inputs to produce s outputs, where the *input-output vector* for farm i is defined as: $(x_i, y_i) \in R_+^{r+s}$. We let x_i^k denote farm i's consumption of the k'th input and y_i^l its production of the l'th output.

The applied DEA output efficiency score under variable returns to scale c.f. [28] for farm i is called θ_i^* and is defined as:

$$\begin{aligned} \theta_i^* = \; & \max \theta_i \\ \text{s.t.} & \sum_{j=1}^{n} \lambda_j x_j^k \leq x_i^k, k = 1, \ldots, r \\ & \sum_{j=1}^{n} \lambda_j y_j^l \geq \theta_i y_i^l, l = 1, \ldots, s \\ & \sum_{j=1}^{n} \lambda_j = 1, \lambda_j \geq 0, j = 1, \ldots, n. \end{aligned} \tag{9.2}$$

The interpretation is that the efficiency score of farm i is the largest possible factor by which farm i can expand all outputs, while still maintaining present input consumption. For further details on the use of DEA, see e.g. [29] or [30].

The resulting benchmarking scores from approximately 7500 farmers function as the additional analyses to help the banks identify the best performing farmers among the many who have too much debt. Table 9.1 shows how the benchmarking scores are distributed within segments of the farmers' debt/equity ratios. The result shows that the vast majority have a debt/equity ratio larger than 50%, and that farmers with similar debt/equity ratio have widely spread benchmarking scores. However, there is a tendency towards a higher benchmarking score for farmers with higher debt/equity scores, i.e. among the most risk-prone customers seen from the banks perspective. Although traditional credit scoring involves elements other than those captured by the debt/equity ratio, the results indicate that the suggested benchmarking analysis is

Table 9.1 Debt/equity ratio and distribution of benchmark scores

Debt/equity ratio (%)	Number of farmers	Average score (%)	Standard deviation (score) (%)
50–60	632	41.6	20.6
60–70	1363	40.0	18.6
70–80	2014	43.0	17.2
80–90	1807	47.8	15.6
90–100	1234	48.3	15.1

able to identify the better performing farmers among the many risk-prone customers who have too much debt.

In conclusion, the confidential benchmarking system was successfully tested by Danish banks as a tool to evaluate potential new customers, as well as groups of farmers in their existing portfolios.

9.3.1.2 Confidential Collaborative Statistics in Finance

This section presents a pilot application of MPC under construction within the Danish R&D project “Big Data by Security”.[9] The application was developed in close collaboration with Danish banks, e.g. Danske Bank and the Danish Growth Foundation. It functions as a Proof-of-Concept of how MPC as a “neutral coordinator” facilitates a new type of collaborative statistics.

Some of the fastest-growing companies have reduced the scope of their business to merely “coordination” via their own platform; a move which has disrupted traditional businesses. Examples of such companies include Airbnb, Uber and crowdfunding operations like Prosper and Lending Club. The scalable business and the information gathered empower these companies. MPC can function as a “neutral coordinator”, the aim of which is to achieve pure coordination between existing data and analytical resources. Therefore, MPC may introduce a new type of disruption that also empowers established companies, which are typically the disrupted entities. This case of privacy-preserving collaborative statistics across banks and banks’ customers provides an example of this.

For incumbents in the financial sector, improved analytics proactively addresses disruption and competition from global players such as Google or Facebook, which have already entered the financial market with digital payments. For new entrants to the financial sector, such as crowd services, privacy-preserving collaborative analytics enhance competition in financial analytics and help entrants keeping a reduced scope of their business, e.g. by buying credit scoring as-a-Service. By opening up the possibility for data collaboration with highly sensitive data, MPC may foster/encourage disruption by making it more valuable to obtain good data from direct sources (e.g. actual transactions), as opposed to huge data from indirect sources (e.g. digital footprints).

The MPC system under construction consists of a generic data module and a number of statistical applications that collectively make up the collaborative statistical service as illustrated in Figs. 9.6 and 9.7.

The data module allows an organizer to orchestrate a data collection job, where a number of providers supply data from independent sources. The data is encrypted when it leaves the providers computer and remains encrypted at all times. All conflicts involved in merging data from different sources are solved without decrypting the data. The consolidated and encrypted data can be used in concrete applications that are specifically tailored and designed for the relevant end users. The data provider

[9] https://bigdatabysecurity.dk.

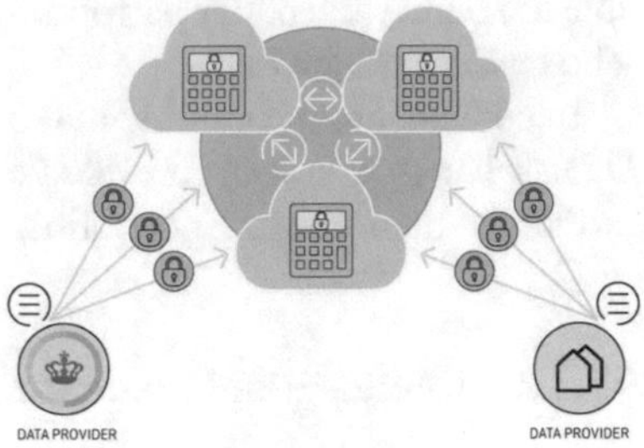

Fig. 9.6 MPC based collection of data

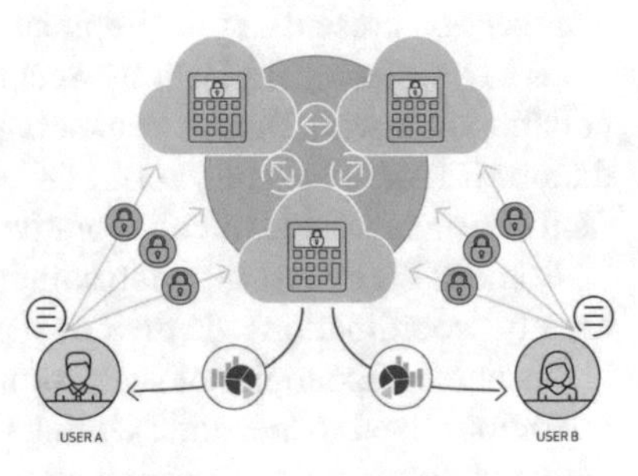

Fig. 9.7 MPC based use of data

knows the intended analyses in advance and accepts the usage when entering data through the data module.

The solution for collaborative statistics is illustrated in Fig. 9.8. The service focuses on risk assessment by providing statistics in the following ways:

- To help the customers select the most relevant/appropriate banks and financial products: The service provides the banks' customers, both new and existing with statistics about their relative performance, as well as credit scoring.
- To help the banks assess existing and potential customers as well as their entire portfolios: The service provides the banks with relative performance data and credit scoring about existing and prospective customers, and their portfolios.

All analyses are based on data gathered from the participating banks, and all results include relative measures from among similar companies, but across all banks.

In addition to the confidential sharing of data, the banks also submit encrypted credit scoring functions. These could be any estimated functions. In the initial use case, the banks provide traditional, discrete versions of the so-called "probability of default" credit scoring functions, although it could be all kinds of regression and machine learning models.

The data and the credit scoring functions are encrypted at all times, and only the results are returned to the relevant users, which are the banks (portfolio analysis) or the banks' customers (individual analysis).

This type of collaborative statistics may result in new types of statistics and insights into the users. Two examples could be:

- To a new bank customer (an outsider): You are among the 50% best scoring customers in Bank A with Bank As own credit scoring function.

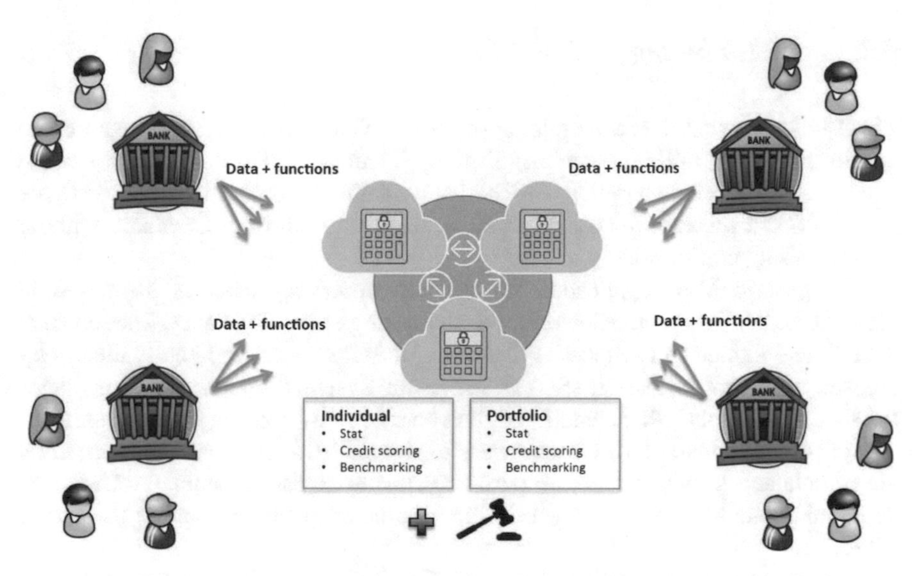

Fig. 9.8 Collaborative statistics among banks and the banks' customers

- To a bank: Your portfolio is 10% better than the market measured with your own credit scoring function.

While privacy-preserving data collaboration itself improves the quality of traditional statistics, the examples above represent new insights that would not be possible without confidential data collaboration.

For a bank customer, a greater understanding of how different banks view them from a statistical point of view, may help in selecting the most appropriate bank and type of loan. To further enhance the competition on the loans market, the solution also includes an auction that makes it possible to get the participating banks to bid for future loans. The collaborative statistics, including the banks' credit scoring functions, make it possible to tailor the description of the supply (the call-off) to the individual bank, which makes it easier for the bank to bid for future loans, which reduces competition in general.

The pilot application is generic and can be extended to many banks and other types of analysis. The solution is suited to the application of machine learning models similar to the confidential use of the banks credit scoring functions. Furthermore, testing and comparing machine learning models across banks would be a relevant extension in the short term. Training machine learning models on encrypted data requires more research, or different machine learning methods.[10]

[10]One example of an ongoing R&D project that focuses on MPC and machine learning is the EU-project SODA (https://www.soda-project.eu).

9.3.2 *On Including National Statistics*

This section presents a second pilot application of MPC that is under development within the Danish R&D project “Big Data by Security”.[11] The application is being developed in close collaboration with Statistics Denmark. Its purpose is to investigate how MPC can increase the use of sensitive data from Statistics Denmark, without compromising confidentiality.[12]

The generic MPC application for privacy-preserving statistics illustrated in Figs. 9.6 and 9.7 has been tailored for benchmarking energy efficiency. The sensitive data were obtained from Statistics Denmark. It includes data on hourly electricity consumption from the Danish TSO (Transmission System Operator), which manages the so-called Datahub that contains all data on electricity consumption in Denmark.

Statistics Denmark allows researchers to run statistics on sensitive information via remote access within a secure perimeter inside Statistics Denmark. The MPC application has been deployed within this perimeter to further explore the use of MPC.

An extended use case could be to use MPC to merge sensitive data across existing public databases, e.g. health data from the Danish Health Data Organisation and socio-economic background data from Statistics Denmark. Consider an MPC system with two computing parties, each controlled by one of two public organizations. The MPC system can merge the data and produce anonymous aggregated statistics. Since neither of the organizations can access the encrypted data, compromising the system would require breaches at both institutions, which is also the case today without any collaboration.[13] In other words, MPC may provide an opportunity for greater data collaboration between various public and private data silos.

High quality data, such as socio-economic data from national statistics, or health-care data obtained directly from doctors and hospitals, could add value to many types of data services. One example could be to use MPC to allow these data to be used to frequently reality test machine learning predictions. An illustrative example is Google Flu Trend that made the headlines when the prediction failed significantly, following a long period of very accurate predictions.[14] Google Flu Trend uses peoples Google search queries to predict future flu epidemics. It turns out that, even before people become sick, many use Google search to find a likely diagnosis. These search queries typically include words related to physical symptoms such as fever and cough”, which are ideal for predicting flu epidemics. However, the reason for

---

[11] https://bigdatabysecurity.dk.

[12] This work has a direct link to the final report from the Commission on Evidence-Based Policymaking in the US. This report explores how to increase the availability of data for assessing government programs, without compromising privacy and confidentiality. The report explains how MPC can be part of the solution as currently tested by Statistics Denmark.

[13] Furthermore, if the MPC protocol is based on so-called secret sharing, as it is, there is no decryption key that can be broken by brute force, which within the EU means that the GDPR regulation may not apply. This is, however, to be legally assessed by the national data protection agencies, who are the legal interpreters of the GDPR.

[14] See e.g. http://gking.harvard.edu/files/gking/files/0314policyforumff.pdf.

the very significant failures was a modification made to the Google search machine, so that it provided "suggested additional search terms" to the user. Following this modification, a large number of people used the "suggested additional search terms", and the prediction went completely wrong. Later, when the problem was discovered, Google fixed the prediction model and Google Flu Trend supplied sound predictions once more. The point is that no one appreciated the problem, and that unforeseen changes may cause automatic prediction models to fail significantly. In this case, MPC could be used to continually test the prediction model using real-time sensitive healthcare data without compromising confidentiality.

9.4 Policy and Managerial Implications

This section discusses public and private policy and managerial implications arising from the use of MPC.

The value added from using MPC is linked to the disruption of the conventional perception of a trustee as an individual or organisation.[15] With no single point of trust, the creation of future "trustees" may significantly change. Also, the fact that confidential information can be shared and used without trust in any single individual or organisation, may have significant disruptive consequences for digitalization in general. Today, the most powerful organizations, such as Google and Facebook, and many others, leverage their power through the tremendous amount of very detailed data they possess on almost all of us. If future cryptographic solutions ensure that individuals and firms can control their private digital information, and if processing is only allowed in encrypted form, the power structures in the digital industry may change.

Today, MPC is a CoED technology on the rise, and the potential disruptive effect has not transpired. However, the use cases presented in this chapter are examples of the changes that may follow these new technical opportunities. Although computational performance prevents some uses today, continued improvements may pave the way for more generic use and substantial disruption.

MPC is a flexible tool and a number of policy and managerial implications may arise from its use and configuration. To see this, remember that MPC is a network of servers (the **CP**s) that collectively constitutes a trustee, where both integrity and confidentiality cannot be compromised by any coalition of t out of the n **CP**s. This basic trust institution can be seen as a tool for creating trust among public and private parties as exemplified below:

- Two or more public institutions could collectively provide a trustworthy institution to private citizens or firms, where no single public institution would have unilateral access to or use of the private data.

[15]A dictionary or a legislative text typically defines a trustee as an individual or institution in a position of trust.

- Two or more competing private entities could collectively establish their own trustee in order to solve a strategic problem such as an auction or a matching based on private information. In the most secure configuration of MPC, with $t = n - 1$, each of the n participants holds the decisive share to access or use the private data, and no one needs to trust the others.

As the two examples illustrate, the MPC systems rules may prevent access to, or use of data, when no consensus exists between $t + 1$ appointed **CP**s. In other words, as if a court order requires 2 out of 3 judges to reveal information, but now built into a cryptographic protocol. Following on from this, one of the basic policy and managerial decisions is to appoint the n **CP**s and decide on a threshold t.

From the point of view of the public, MPC can be used to create trust in public use of sensitive data. Trust is probably one of the more important ingredients required for a well-functioning society. Enhanced trust may open up the potential for increased use of sensitive data to make valuable data-driven decisions for the greater good. The case of Statistics Denmark illustrates this. Other countries may gain even more from greater governmental trust. The report by "Commission on Evidence-Based Policymaking" in the US, mentioned in Sect. 9.3.2 may be an indication of this.

From a private point of view, MPC can be used to build trust between competitors, as illustrated above. The first commercial uses of MPC have done exactly that by e.g. replacing the auctioneer, or the trustee, on off-exchange financial matching. The financial use case on collaborative statistics presented in Sect. 9.3.1 is another, but more advanced example of how MPC facilitates collaboration between competitors.

In both cases, MPC reduces the cost of creating trust. A traditional trustee is costly as it relies on reputation and strict rules of conduct for individuals, among other things. MPC reduces costs by building on existing trust or opposing interests through the appointed **CP**s. As the MPC system does not rely on individuals or institutions, the risk of human error when handling confidential data is reduced to a minimum.

From a public regulatory perspective, MPC is a tool that can ensure the continued use of sensitive personal information, despite data regulation such as GDPR. MPC may also be an important means for competition authorities to control the use of sensitive information, as opposed to the prevention of sharing the same information. To give an example, information about storage capacity in a supply chain may be used to coordinate prices, but it may also reduce over-production and waste. With a high level of informational control, the competition authorities may allow the confidential collection of information on storage utilization if only used for statistical purposes that directly relate to reducing over-production and waste. To what extent a statistical computation can be isolated for a specific purpose, such as waste management, requires individual judgement. Therefore, the future use of MPC may change the role of regulators from approving the sharing of data, to approving computations and results.

9.5 Discussion and Prospects

Big Data and the correct analytics have tremendous value, some of which is created by using sensitive personal or company data as raw inputs. On the one hand, data breaches and some uses of these sensitive data may violate fundamental rights to privacy, as well as strategic information about preferences and bargaining positions. On the other hand, the use of these sensitive data may solve fundamental problems in fraud detection, risk assessment, testing of medical treatments, and tailored health-care services and may lead to the improvement of numerous other services for the greater good.

As a buyer, it may be quite sensible to use a service which can reduce searching and matching costs, and which makes it easier and more user-friendly to find the right match to the buyer's preferences. However, if the service reveals the buyer's willingness to pay in the process, and if the real income stream for the service is to sell the revealed preference information on a B2B market for Big Data insights, then it might be better to avoid using the service in the first place. One could say that if you do not explicitly pay for confidentiality, one should expect the revealed information to be the payment. One conclusion of this could be that the market solves the problem, and those who value their private information pay for it, while the remainder pay with their private information as a currency.

In practice, both the harvesting of Big Data (such as digital footprints, transactions and IoT sensors data), and its use are not transparent. Therefore, another conclusion may be that regulation is required to ensure that everyone is fully informed about the use of sensitive data, or to prevent use of the same, or something in between. The GDPR is the most progressive regulation of personal identifiable information and will come into effect in May 2018. Therefore, the effect of this type of regulation is unclear at this stage and will rely on the legal interpretation by the national data protection agencies within the EU.

Whether confidential use of sensitive data is driven by market demand or is pushed through by regulation, the market needs technical solutions that can provide the desired level of security and privacy in a convenient and user-friendly way.

One solution is to make data anonymous, which might be sufficient for some types of aggregated statistics. However, any anonymization technically degrades the value of the data, which is a particular problem in services such as risk assessment and fraud detection that mostly rely on actual sensitive data. As described in this chapter, CoED techniques can solve this problem and allow computation on the actual, albeit encrypted data.

In general, strong encryption is a fundamentally sound way of protecting data, and solutions for securing data "in transit" and "at rest" are standardized commercial products. These solutions are also simple for users to monitor through, e.g. standards such as SSL connections and certificates. The missing part is a generic standardized and certified Computation on Encrypted Data (CoED). Although MPC and other CoED techniques can be used for any generic computation, the computational requirements prevent their generic use today. However, as presented in this chapter, the performance of CoED has improved exponentially, and the uptake in various commercial solutions has started.

MPC is a CoED technique that does not rely on trusted individuals or institutions. Therefore, MPC can be used to create more trustworthy institutions than those we have today. One solution is to use MPC to combine existing trusted institutions to create one that is more neutral and trustworthy. Another solution is to use opposing interests to further increase the trustworthiness of the MPC.

MPC has been successfully used commercially within market solutions such as auctions and matching, and also basic secure infrastructure, such as key management. Commercial use of MPC within statistics is probably the next field of use. As the many prototypes and pilot applications indicate, MPC is sufficiently efficient for privacy-preserving statistics.

The special security feature of MPC; that it does not rely on a trusted third party, opens up new opportunities for data collaboration, such as the case with the Danish banks and their customers. With MPC as the trustworthy and neutral coordinator, competing banks can pool their data and produce statistics that benefit all banks, and allow new services to be offered to the banks' customers. The setup may also result in other services, such as credit-scoring-as-a-Service, to be made available to emerging banking competitors, such as crowdfunding.

Acknowledgements This research has been partially supported by the EU through the FP7 project PRACTICE and by the Danish Industry Foundation through the project "Big Data by Security".

References

1. H. Varian, Economic mechanism design for computerized agents, in *First USENIX Workshop on Electronic Commerce* (1995)
2. V. Costan, S. Devadas, *Intel SGX Explained* (Cryptology ePrint Archive, Report 2016/086)
3. A. Gibbard, Manipulation of voting schemes: a general result. Econometrica **41**, 587–601 (1973)
4. R.B. Myerson, Incentives compatibility and the bargaining Problem. Econometrica **47**, 61–73 (1979)
5. A. Shamir, How to share a secret. Commun. ACM **22**(11), 612–613 (1979)
6. D. Chaum, C. Crepeau, I.B. Damgaard, Multiparty unconditionally secure protocols (extended abstract), in *20th ACM STOC, Chicago, Illinois, USA, 24 May 1988* (ACM Press, 1988) pp. 11–19
7. P. Bogetoft, I.B. Damgaard, T. Jacobsen, K. Nielsen, J.I. Pagter, T. Toft (2005), Secure computation, economy, and trust—A generic solution for secure auctions with real-world applications. Basic Research in Computer Science Report RS-05-18
8. P. Bogetoft, D.L. Christensen, I.B. Damgaard, M. Geisler, T. Jakobsen, M. Krigaard, J.D. Nielsen, J.B. Nielsen, K. Nielsen, J. Pagter, M.I. Schwartzbach, T. Toft, Secure Multiparty Computation Goes Live. Lecture Notes in Computer Science **5628**, 325–343 (2009)
9. D. Malkhi, N. Nisan, B. Pinkas, Y. Sella, Fairplay—A secure two-party computation system, in *Proceedings of the 13th USENIX Security Symposium* (2004), pp. 287–302
10. P. Bogetoft, K. Boye, H. Neergaard-Petersen, K. Nielsen, Reallocating sugar beet contracts: can sugar production survive in Denmark? Eur. Rev. Agric. Econ. **34**(1), 1–20 (2007)

11. B. Pinkas, T. Schneider, N.P. Smart, S.C. Williams, *Secure Two-Party Computation Is Practical* (Asiacrypt, 2009)
12. A. Shelat, C. Shen, *Two-output Secure Computation With Malicious Adversaries* (Euroscript, 2011)
13. J.B. Nielsen, P.S. Nordholt, C. Orlandi, S.S. Burra, *A New Approach to Practical Active-Secure Two-Party Computation* (Crypto, 2012)
14. T.K. Frederiksen, J.B. Nielsen, *Fast and Maliciously Secure Two-Party Computation Using the GPU* (ACNS, 2013)
15. T.K. Frederiksen, J.B. Nielsen, *Faster Maliciously Secure Two-Party Computation Using the GPU* (SCN, 2014)
16. Y. Lindell, B. Riva, *Blazing Fast 2PC in the Offline/Online Setting with Security for Malicious Adversaries* (CCS, 2015)
17. B.N. Nielsen, T. Schneider, R. Trifiletti, *Constant Round Maliciously Secure 2PC with Function-independent Preprocessing using LEGO* (NDSS, 2017)
18. I. Damgaard, K.L. Damgaard, K. Nielsen, P.S. Nordholt, T. Toft, Confidential benchmarking based on multiparty computation. *Financial Cryptography and Data Security*. Lecture Notes in Computer Science, vol. 9603 (Springer, 2017), pp. 169–187
19. A. Charnes, W.W. Cooper, E. Rhodes, Measuring the efficiency of decision making units. Eur. J. Oper. Res. **2**, 429–444 (1978)
20. A. Charnes, W.W. Cooper, E. Rhodes, Short communication: measuring the efficiency of decision making units. Eur. J. Oper. Res. **3**, 339 (1979)
21. E. Emrouznejad, B.R. Parker, G. Tavares, *Evaluation of research in efficiency and productivity: a survey and analysis of the first 30 years of scholarly literature in DEA* (Soc. Econ. Plann, Sci, 2008)
22. P.J. Agrell, P. Bogetoft, J. Tind, DEA and dynamic yardstick competition in Scandinavian electricity distribution. J. Prod. Anal. **23**(2), 173–201 (2005)
23. P. Bogetoft, K. Nielsen, DEA based auctions. Eur. J. Oper. Res. **184**(2), 685–700 (2008)
24. K. Nielsen, T. Toft, Secure relative performance scheme, in *Proceedings of Workshop on Internet and Network Economics*. LNCS (2007), 48–58
25. A. Cielen, L. Peeters, K. Vanhoof, Bankruptcy prediction using a data envelopment analysis. Eur. J. Oper. Res. **154**(2), 526–532 (2004)
26. J.C. Paradi, M. Asmild, P.C. Simak, Using DEA and worst practice DEA in credit risk evaluation. J. Prod. Anal. **21**(2), 153–165 (2004)
27. I.M. Premachandra, G.S. Bhabra, T. Sueyoshi, DEA as a tool for bankruptcy assessment: a comparative study with logistic regression technique. Eur. J. Oper. Res. **193**(2), 412–424 (2009)
28. R.D. Banker, A. Charnes, W.W. Cooper, Some models for estimating technical and scale inefficiencies in data envelopment analysis. Manage. Sci. **30**, 1078–1092 (1984)
29. P. Bogetoft, L. Otto, *Benchmarking with DEA, SFA, and R* (Springer, New York, 2011)
30. W.W. Cooper, L.M. Seiford, K. Tone, Data envelopment analysis: a comprehensive text with models, applications, references and DEA-solver software, 2nd edn. (Springer, 2007)

Kurt Nielsen has contributed to theory and real-life applications of market design, decision support systems and information systems in broad terms. As researcher he has co-initiated several cross-disciplinary research projects involving computer science, economics and industry partners. This work has resulted in a variety of academic publications across economics and computer science. As entrepreneur, he has co-founded four startups and been involved in numerous software projects that bring state-of-the-art research together with business applications as well as critical business solutions such as governmental spectrum auctions and confidential management of sensitive data.